Deep Injection Disposal of Liquid Radioactive Waste in Russia

Andrei Ivanovich Rybal'chenko
Mikhail Koz'mich Pimenov
Petr Petrovich Kostin
Valentina Dmitrievna Balukova
Anatolii Viktorovich Nosukhin
Evgenii Il'ich Mikerin
Nikolai Nikolaevich Egorov
Elena Pavlovna Kaimin
Inessa Mikhailovna Kosareva
Vitalii Mikhailovich Kurochkin

Edited by:
Michael G. Foley
Lisa M. G. Ballou

Translated by:
Ben Teague

 Battelle Press

Columbus • Richland

Disclaimer

This report was prepared as an account of work sponsored by an agency of the United States Government. Neither the United States Government nor any agency thereof, nor Battelle Memorial Institute, nor any of their employees, makes any warranty, express or implied, or assumes any legal liability or responsibility for the accuracy, completeness, or usefulness of any information, apparatus, product, or process disclosed, or represents that its use would not infringe privately owned rights. Reference herein to any specific commercial product, process, or service by trade name, trademark, manufacturer, or otherwise does not necessarily constitute or imply its endorsement, recommendation, or favoring by the United States Government or any agency thereof, or Battelle Memorial Institute. The views and opinions of authors expressed herein do not necessarily state or reflect those of the United States Government or any agency thereof.

PACIFIC NORTHWEST NATIONAL LABORATORY
operated by
BATTELLE
for the
UNITED STATES DEPARTMENT OF ENERGY
under Contract DE-AC06-76RLO 1830

Library of Congress Cataloging-in-Publication Data

Glubinnoe zakhoronenie zhidkikh radioaktivnykh otkhodov.
English. Deep injection disposal of liquid radioactive waste in Russia/
    Rybal'chenko Andrei Ivanovich;
    translated by Ben Teague;
    edited by Michael G. Foley, Lisa M. G. Ballou.
    p.   cm
    Includes bibliographical references (p.).
    ISBN 1-57477-064-0 (alk. paper)
    1. Radioactive waste disposal in the ground—Russia (Federation)
    2. Deep-well disposal—Russia (Federation)
I. Rybal'chenko, Andrei Ivanovich. II. Foley, Michael G.   III. Ballou, Lisa M. G.   IV. Title
TD815.4.R8G5513 1998
621.48'38—dc21   98-25116
                                                                                CIP

Printed in the United States of America

Battelle Press, 505 King Avenue, Columbus, Ohio 43201-2693, USA
614-424-6393 or 1-800-451-3543. Fax: 614-424-3819, E-mail: press@battelle.org
Website: www.battelle.org/bookstore

Editors' Preface

This book, published by Izdat in Moscow, Russia in 1994, tells the story of the first 40 years of work in the former Soviet Union to devise, test, and execute a program to dispose of by deep injection millions of cubic meters of liquid radioactive wastes from nuclear materials reprocessing. The main focus of the book, planned as the first of two books on radioactive waste injection in Russia, is to explain past decisions regarding deep injection disposal of liquid radioactive wastes resulting from the former Soviet Union's Cold War weapons production activities. Those decisions involved safety, research, and practical experience gained during creation and operation of disposal systems. The deep injection of wastes was accompanied by systematic monitoring through a network of instrumented wells, together with hydrogeologic and radiochemical analyses. The results of these continuing studies, now pursued jointly by the Ministry of Atomic Energy of the Russian Federation (MINATOM) and the U.S. Department of Energy (DOE), will be the subject of the planned second book on deep injection of liquid radioactive wastes in Russia.

The handling of radioactive wastes is one of the world's most controversial topics. The authors of this book, all of whom participated in development of the former Soviet Union's waste injection program, considered that the optimal development and use of nuclear power in Russia to stabilize the economy and fortify national defense might be jeopardized. Accordingly, one objective for publishing this book in Russia was to set forth the results of studies and practical efforts in a form accessible to a broad audience. The authors made this book available to Dr. Frank Zanner of Sandia National Laboratories (SNL); Ben Teague translated the Russian version for SNL. At the same time, Dr. Nikolai Egorov, Deputy Minister of MINATOM and one of the book's authors, requested that DOE publish the English translation of the book.

Under the auspices of the DOE-MINATOM Joint Coordinating Committee for Environmental Management (JCCEM), Pacific Northwest National Laboratory (PNNL) has been developing jointly with MINATOM scientists three-dimensional numerical models of the hydrogeology and potential contaminant migration at the two main radioactive waste injection sites in the West Siberian Basin since 1991. Given the difficult nature of technical translation and interpretation, Dr. Clyde W. Frank of DOE, the JCCEM coordinator, requested that PNNL edit Mr. Teague's translation using their detailed knowledge of the sites and their injection histories, and their access to the Russian authors, then arrange for publication in the U.S. Dr. Michael Foley, the PNNL project manager and technical leader for the West Siberian Basin studies, provided technical interpretation to ensure that the translation correctly reflects the Russian concepts in Western terminology. Ms. Lisa Ballou, who has been the principal technical editor/writer for PNNL's West Siberian Basin studies since their inception, performed the difficult task of converting the translation into conventional English syntax, designed the book format and layout, and supervised preparation of the final copy. Mr. Joe Sheldrick of Battelle Press provided invaluable assistance in preparing the book for publication.

Preface

The deep injection disposal (burial) of liquid radioactive wastes has been practiced for 30 years. Some 46 million m^3 of wastes, containing over half the radioactive nuclides (fission products of uranium) generated as waste in Russia's nuclear industry, have been placed in deep reservoir formations. Consequently, industrial workers and the general population near two defense establishments and a large research center have been protected from the detrimental effects of radioactive wastes.

Injection disposal sites (burial sites) for liquid radioactive wastes were created in a fairly short time on the basis of comprehensive studies and development work involving organizations belonging to several agencies, including Minatom, the Academy of Sciences, the Ministry of Geology, and the Ministry of Public Health. Observations and studies carried out during the operation of these disposal sites have confirmed the effectiveness of isolating radioactive wastes in deep reservoirs and the soundness of the basic decisions made when the sites were created.

At the same time, the deep injection disposal of liquid radioactive wastes is now a subject of debate as the consequences of defense activities in the nuclear industry are discussed. There are many instances of subjective and unduly negative assessments by ecology movements, the mass media, legislative bodies, and government representatives. As a result, poor decisions may ensue and affect the future work of enterprises as well as the operation and deactivation of disposal sites.

This situation results partly from the lack of adequate information on the technology of deep injection disposal of liquid radioactive wastes. This book, therefore, discusses the reasons for setting up disposal sites and the requirements for suitable geological formations, describes practical experience as well as research results from site operation, and offers an evaluation of safety and the consequences of radioactive waste burial.

In this book, the material and its organization are oriented toward providing exhaustive answers to those questions that arise when waste disposal and environmental protection are discussed by representatives of society, the government, and deputies in the legislature. This book will also appeal to geologists, mining engineers, environmental protection experts, and anyone who takes a professional or personal interest in waste handling and the use of geological formations for disposal purposes.

Acknowledgments

For the research and development work described in this book, we acknowledge the agencies of the Ministry of Geology, above all the Production State Association Hydrospetzgeologiya. Their contributions were valuable for solving the complex and important problem of deep injection disposal. Chief among the researchers were B. N. Savvin, S. T.Gavrilov, M. M. Polyakov, A. A. Goncharov, I. I. Tishchenko, A. V. Nosukhin, A. T. Larchenko, B. Ya. Lebedev, V. T. Ryzhenkov, G. P. Popsui-Shapko, and several specialists in the various plant geological services: M. N. Baranov, V. P. Salopov, L. F. Novoselov, B. P. Sigaev, I. A. Lachkov, T. N. Nosukhina, V. M. Ovchinnikov, I. N. Povedailo, R. V. Pershina, A. V. Mitryushin, A. S. Ladzin, and I. P. Ulyushkin.

The problem of deep disposal of liquid radioactive wastes could not have been solved without the fruitful work of scientists and designers at the institutes of the Ministry of Medium Machinery Construction (Minatom at present) and the Academy of Sciences. Making great contributions were (along with authors of this book) F. P. Yudin, A. N. Kalinin, E. N. Munaev, V. G. Panasyuk, D. I. Levitskii, Yu. I. Abramov, G. A. Okun'kov, G. P. Vakhurina, A. A. Menyailo, E. V. Zakharova, A. M. Rykov, and M. L. Medvedeva.

All these efforts were supported by plant directors and leading figures in the Ministry: A. G. Meshkov, E. I. Mikerin, A. D. Zverev, N. I. Grekov, N. N. Egorov, A. R. Belov, O. D. Kazachkovskii, S. I. Zaitsev, N. S. Osipov, A. I. Karelin, L. M. Khasanov, and others. Major aid was provided by staffers in agencies of the Ministry: B. S. Kolychev, Z. V. Chausov, A. N. Tymanyuk, and N. A. Rakov.

Introduction

Prevention of the harmful impact of radioactive wastes on human beings is an urgent problem for the second half of the twentieth century, attracting the attention of scientists and politicians, as well as broad sections of society. The danger of increased radiation levels became quite clear after the use of nuclear weapons by the U.S. in 1945 and later in research on the production of fissionable materials. As a result, a special relationship to atomic energy and radioactive wastes came about, requiring both fundamental and immediate measures to isolate such wastes from human beings.

Special hazards were posed by liquid radioactive wastes, which were generated in large quantities and contained the bulk of the radioactive nuclides produced by the fission of uranium. The hundreds of thousands of cubic meters of liquid wastes that were generated yearly were difficult to localize; wastes had contaminated surface waters and large land areas.

The urgent nature of the nuclear programs in the USSR and the U.S. demanded that certain engineering problems be solved ahead of disposal problems. As a result, in the late 1940s and early 1950s, issues regarding handling of radioactive wastes and environmental protection were pushed aside. In addition, at the time there was a lack of knowledge about many aspects of radionuclide behavior in the environment.

When radiochemical plants were just coming into service, liquid radioactive wastes were handled in the same way as ordinary industrial wastes. Experience proved how unacceptable this approach was. Land adjacent to exposed lagoons and liquid radioactive waste storage sites was contaminated by aerosols; entrainment by winds proved to have extremely serious consequences when cyclones occurred. Seepage from surface reservoirs contaminated groundwaters, and streams including major rivers were contaminated for hundreds or thousands of kilometers away from plants.

The storage of liquid radioactive wastes in special vessels also proved to be dangerous, as was shown by the 1957 explosion of a surface reservoir and resulting contamination of large tracts of the southern Urals. In other instances, nuclides were observed in ecological chains throughout the regions of radiochemical plants, and biomigration of nuclides was also found. Surface storage of liquid radioactive wastes presented a great hazard in case of military attack, where the scale and level of contamination by radioactive wastes would be far greater than would result even from an atomic explosion. These facts impelled the Soviet and

The urgent nature of the nuclear programs in the USSR and the U.S. demanded that certain engineering problems be solved ahead of disposal problems.

American governments to put scientists and specialists to work looking for solutions to the radioactive waste problem.

In the USSR, analysis of radioactive waste accidents and consultation with leading specialists in the atomic industry and at the Academy of Sciences revealed three main approaches to dealing with radioactive wastes: 1) processing to recover the nuclides and other substances of interest for a variety of subsequent uses; 2) conversion of wastes to a solid form followed by storage and burial; and 3) disposal of wastes, including liquid wastes, in geological formations immediately following waste generation.

The use of geological formations as the final destination for liquid radioactive waste was not by chance. Many types of deep geological formations lie outside the sphere of human activity and are not involved in the rapid circulation of living material, which takes place chiefly in the zone where the Earth's surface and the atmosphere come together. Deep geological formations are not easily accessible to accidental or deliberate penetration.

Some formations are, in a manner of speaking, the "wastes" of former natural biosystems. These include coal beds (naturally buried and transformed remains of ancient plants), petroleum, natural gas, bitumens (also organic residues according to one hypothesis), phosphate rocks, and oolitic limestones. Where conditions were favorable, these deposits of former wastes, formed hundreds of thousands or millions of years ago, have been preserved down to our time. Most of them do not appear at the surface and can only be identified with difficulty in surveys conducted for just this purpose.

The formation and burial of such wastes accompanied the development of all formerly existing biosystems. The wastes in turn played a definite role in the existence and development of later biological communities. This phenomenon is still typical as deposits of coal, oil, and gas are rapidly used by humans.

The processes by which waste products are generated, accumulated, and used are self-regulated and involve the natural creation and circulation of living material. The controls on such processes can, however, be disrupted by the adoption of certain paradigms. For example, human needs must be maximally satisfied no matter what the cost (where these needs are often wrongly understood and incorrectly ranked), or humanity must make every effort to transform nature. Neither of these paradigms takes into account the complexity and interrelationship of natural processes and conditions as a whole, of which the human race and society are just a part. The arms race, the striving for superiority or parity in the military arena at all costs, exacerbated the negative consequences of scientific and technical progress. The restoration of a natural equilibrium upset by human activity—including the generation of wastes—is a very urgent task for the present and the near future.

Proposals for the disposal of liquid radioactive wastes in geological formations have been advanced by geologists, radiochemists, and petroleum specialists, among them Academicians A. P. Vinogradov and V. I. Spitsin, Professor S. A. Voznesenskii, and N. A. Kalinin. Their ideas were based on existing knowledge and practical experience in exploration, prospecting, and development of oil and gas fields and other mineral deposits. It was well known that oil and gas pools survive for hundreds of thousands or millions of years beneath "caps" of relatively impermeable rocks and that sometimes, at high ("abnormal") formation pressures, pools can become part of tectonic zones. Some deposits of metals, including uranium (supergene deposits), were formed by the transfer of material from groundwaters to rocks through which metal-laden waters flowed. In other words, the creation of these deposits amounted to natural purification of groundwaters. The relative stability of deep geological formations over time and the difficulty of accessing them, either accidentally or deliberately, were also important factors arguing for underground disposal of liquid radioactive wastes.

Preliminary discussion of the underground disposal of liquid radioactive wastes led to the realization that all geological formations by no means meet the requirements for isolating liquid radioactive wastes. For this reason, disposal must always be preceded by investigation of the geological structure and properties of the rocks, the wastes, and their compatibility with the geological surroundings.

On the strength of proposals by scientists and specialists, the Soviet government decided in the late 1950s to begin geological explorations and studies aimed at identifying and establishing deep injection disposal sites for liquid radioactive wastes at four nuclear enterprises: the Siberian Chemical Combine (SCC) at Tomsk-7 (now Seversk), the Mining and Chemical Combine (MCC) at Krasnoyarsk-26 (now Zheleznogorsk), the Mayak Production Association at Chelyabinsk-65 (now Ozersk), and the Scientific Research Institute of Nuclear Reactors at Dimitrovgrad (NIIAR) (Ul'yanovsk oblast).

In view of the increasing impact of radioactive wastes on the environment, a rush schedule was established for the development of disposal sites. Initially, deep injection disposal of liquid radioactive wastes was thought of as a temporary measure until technologies could be devised for reprocessing and solidifying the wastes, and although broad-based research was instituted to this end, it took far longer to complete. It was assumed that converting the wastes to solid forms and simultaneously reducing their physical volume would offer great freedom in choosing a

All geological formations by no means meet the requirements for isolating liquid radioactive wastes. ...disposal must always be preceded by investigation...

final method of disposal: storing wastes at the surface, burying them in deep, low-permeability formations, firing them into outer space, or reprocessing them, for example.

At the same time, it was plain that volume is a key factor in choosing a disposal method for liquid wastes. While reprocessing and solidification is a fairly efficient technique for the relatively small volumes of wastes from medical institutions and small research centers, a grave problem arises for large radiochemical plants generating from hundreds to thousands of cubic meters of waste each day.

Following analysis of the options for dealing with radioactive wastes, it was determined that disposal of liquid wastes in reservoir formations could be implemented much more quickly than solidification or thorough reprocessing. This timeliness made a crucial argument for deep injection disposal of liquid radioactive wastes, since the accumulation of these substances at the surface presented a major potential hazard. This argument was borne out because solidification technology was not adopted until much later, and waste reprocessing has not been realized even yet. Work on the disposal of liquid radioactive wastes has generally been in accordance with International Atomic Energy Agency (IAEA) Recommendations, which have been periodically re-examined since the second half of the 1950s.[1]

Attempts were made to establish underground disposal sites for liquid radioactive wastes in the U.S., where the deep injection of wastes from the chemical, pharmaceutical, and other industries had been practiced for decades. Analysis showed, however, that the principal operations of the U.S. nuclear industry were located in areas where the geological conditions did not favor deep injection disposal of liquid wastes.[2] Even so, liquid and slurry radioactive wastes from the research center at Oak Ridge, Tennessee were disposed of into low permeability rocks (shales).[3] At the National Reactor Test Facility in Idaho (now Idaho National Engineering and Environmental Laboratory), low-level wastes were injected through boreholes into a formation of jointed rocks; the geological conditions were relatively poor. Later, this test was stopped when large-scale waste migration was detected.[4] Wastes from industrial uranium refining were injected into subsurface formations by Anaconda in New Mexico.

The first geological exploration in the USSR aimed at creating a deep injection disposal site for liquid radioactive wastes was begun in the middle 1950s around the SCC. Carried out by the Novosibirsk Territorial Geological Administration, this project included seismic and electrical surveys, geological mapping, drilling, and flow tests. The conduct of subsequent work was passed to a specialized agency, the All-Union Hydrogeological Trust (later renamed Second Hydrogeological Administration of the Ministry of Geology, now Production State Association Hydrodrospetzgeologiya), in an effort directed by K. I. Antonenko, M. Ya. Danilovich, E. G. Chapovskii, and Yu. S. Tatarchuk. Physical and chemical studies of the wastes, their compatibility with the geological surroundings, and development of a technology to condition wastes to be disposed of were performed by the Academy of Sciences Institute of Physical Chemistry under the direction of Academician V. I. Spitsin and V. D. Balukova. At the All-Russian Scientific Research and Design Institute of Industrial Technology (formerly Promniiproekt), E. D. Mal'tsev, F. P. Yudin, and M. K. Pimenov managed scientific, technical, economic, and design studies, including safety aspects, simulation of burial conditions, and design of engineering systems. For the design of surface facilities, the scientific research and design institute VNIPIET (M. V. Strakhov) was enlisted. Public health issues in the creation of disposal sites for liquid radioactive wastes were handled by the Institute of Biophysics and the third Medical Administration of the Ministry of Public Health under the direction of A. S. Belitskii and A. I. Ryzhov.

In the late 1950s, geological exploration and research was also undertaken in the regions of the MCC, the Mayak Production Association, and the NIIAR. The data led agencies of the Ministry of Geology to conclude that disposal of liquid radioactive wastes was feasible in the regions of the SCC, the MCC, and the NIIAR. Initial design figures were also developed. The geology of the Mayak site was unfavorable for deep injection disposal of the wastes; later it was determined that the nearest place satisfying the essential requirements was 150 km from the plant in the Kurgan oblast. It was decided not to proceed with deep injection disposal at this site.

A complicating factor at the SCC and MCC was the presence of fresh water—a potential resource—in reservoir formations suitable for waste injection. The scientists and specialists took the general view that formations containing saline water would be preferable for waste disposal because they had no other economic value. However, there were at that time no alternative technologies for handling these potentially hazardous liquid radioactive wastes; it was necessary to dispose of them near the plants. Since injection of wastes would remove only small reserves of fresh groundwater from other possible uses, it was decided that the formations identified could be used for the disposal of liquid radioactive wastes.

In accordance with the ratified design documents, an experimental site was created for the disposal of process wastes from the SCC. This was placed in service in 1963 and was followed by sites at Area 18 and Area 18a in 1967 and 1975, respectively. The liquid radioactive waste storage sites at the MCC were put in operation in 1967-69, those at the NIIAR between 1966 and 1973. The deep injection of wastes was accompanied by systematic monitoring through a network of instrumented wells, together with special studies. The results of these measurements are of interest in themselves.

The handling of radioactive wastes is a hotly debated topic; quite often, decisions previously made are examined from a subjective point of view, and the facts are distorted. The optimal development of the nuclear power industry is harmed in this way, and the maximum use of nuclear power to stabilize Russia's economy and fortify national defense is put in question. Accordingly, one objective of this book is to set forth the results of studies and practical efforts in a form accessible to a broad audience.

Results achieved in studies of deep injection disposal are useful in solving other problems involving the economic use of the area beneath the Earth's surface, and the material in this book is, therefore, presented with an eye to possible later applications. Because liquid radioactive wastes are so toxic and the decisions made are so vital, many topics in waste disposal have been examined with special care. Developments in this area will be of great interest for the disposal of nonradioactive wastes. The first specialized interdepartmental standard on underground disposal was created for liquid radioactive wastes but has come to be applied to other categories of industrial waste products.[5]

The main focus of the book is to justify past decisions regarding deep injection disposal of liquid radioactive wastes, including safety aspects, research results, and practical experience gained during creation and operation of disposal systems. The book is organized around results from geological exploration, scientific research, and experience at two large enterprises of the nuclear industry and one research center.

Work done at enterprises where research and design efforts did not, for a variety of reasons, lead to the establishment of disposal sites is discussed only briefly. And a number of topics receive only minimal attention including aspects of mass transport in permeable and relatively impermeable formations, and nonequilibrium chemical and physical processes. Further advances in these areas will not affect the basic safety assessment but will hold interest for geologists, theoreticians, and site management and staff. The copious experimental data on deep injection disposal of liquid radioactive wastes still await researchers.

Contents

Figures

Tables

1 Liquid Radioactive Wastes and Methods of Handling Them

1.1 Classification of Liquid Radioactive Wastes

The term "liquid radioactive wastes" denotes solutions containing radioactive nuclides, which for technical, economic or other reasons cannot be used in the manufacture of a salable product or for other economic purposes. The composition and the chemical and physical properties of liquid radioactive wastes differ widely, depending on the location and conditions under which the wastes were generated, as well as pretreatment prior to forwarding for disposal. The total activity and radionuclide composition are basically dictated by the fission products of uranium. Along with radionuclides, liquid radioactive wastes contain salts, metal ions, acids, alkalies, detergents, organic compounds and many other substances, as well as solids in the form of finely divided suspended matter, colloids, and gels.

The lower bound on the radionuclide contents, above which the wastes are deemed radioactive, is usually taken as the permissible concentrations of the radionuclides in potable water as set forth in radiation safety standards.[6] To evaluate the toxicity of liquid radioactive wastes in geological formations, researchers have suggested that wastes be regarded as nonhazardous if the radiation levels are below the levels existing in natural uranium, thorium and potassium isotopes in the rocks or groundwaters (this limit is applied with allowance for the variability of such contents).[7]

Radioactive wastes, which are usually classified by specific activity (number of radioactive decays per unit time and unit quantity of the substance), fall into three categories, low-level waste, intermediate-level waste, and high-level waste. Experience has shown that such a classification is ambiguous, especially for liquid wastes, since the level of activity is not the only factor governing the biological hazard or toxicity of wastes. But, toxicity is the main criterion for selecting a method of handling wastes—particularly for storage and disposal. For example, intermediate-level wastes containing long-lived transuranic (TRU) nuclides and fission products (^{90}Sr and ^{137}Cs) will be more dangerous in long-term isolation than will high-level wastes containing only short-lived fission products (^{144}Ce, ^{147}Pm, ^{106}Ru, ^{103}Ru, ^{95}Zr, and ^{95}Nb, and tritium). For this reason, the ranges of specific activity used to assign wastes to one category or another will differ widely between nations, and other categories are used in some cases.[8] Other criteria for classification include the half-life of the radionuclides and the type

Toxicity is the main criterion for selecting a method of handling wastes —particularly for storage and disposal.

of radiation, the heat production per unit mass, the conditions of waste formation, and the methods used to handle the wastes.

The International Atomic Energy Agency (IAEA) was the first to attempt to devise a unified classification system for liquid radioactive wastes (in 1970).[9] Low-level wastes comprised three subclasses with activities of $<10^{-6}$ Ci/m^3, from 10^{-6} to 10^{-3} Ci/m^3, and from 10^{-3} to 10^{-1} Ci/m^3; the intermediate class took in activities of 0.1 to 10^4 Ci/m^3; wastes with activities of more than 10^4 Ci/m^3 were considered high level. Many countries employ national classification systems for radioactive wastes, including liquid wastes, taking into account the level of development and the nature of the nuclear technologies practiced, as well as the technical, economic, social conditions, and the natural environment.

In the Russian Federation, radiation safety and public health regulations are essentially contained in the codes NRB-76/87, OSP-72/87, and SPORO-85.[6,10] Under these regulations, liquid radioactive wastes are classified as low level ($< 10^{-5}$ Ci/L), intermediate level (from 10^{-5} to 1 Ci/L) and high level (> 1 Ci/L). In some scientific and technical literature, such as the book *Decontamination of Liquid Radioactive Wastes*,[11] the high-level category is extended to solutions requiring refrigerated storage.

The U.S. has used a two-criteria system for classifying radioactive wastes. The first criterion is the activity level of the wastes, governed chiefly by the concentration of the fission products making up the principal sources of heat and external ionizing radiation. The second criterion is the concentration of radionuclides with very long half-lives, principally alpha emitters that dictate how long the wastes must remain isolated.[8]

There are three main categories of radioactive wastes: high-level wastes, alpha-emitting transuranic wastes, and low-level wastes. High-level wastes feature high activity levels and generation of enough heat that special practices are needed to limit the short-term risk in handling them. High-level wastes may contain very long-lived nuclides and thus require long-term isolation. TRU wastes or equivalent wastes with high concentrations of long-lived nuclides (^{14}C, ^{99}Tc, ^{129}I, etc.) need long-term isolation but do not have the activity of high-level wastes.

The identification of high-level wastes in the U.S. is based on a heat output of over 50 W/m^3 or an external ionizing radiation level of over 1 Sv/hr at a distance of 1 m from the wastes. Concentration limits are established for each radionuclide according to this threshold. The concentration limit for ^{90}Sr, for example, is 7000 Ci/m^3; for ^{137}Cs, 5000 Ci/m^3; and for ^{239}Pu, 2000 Ci/m^3. The limit for a waste containing a mixture of nuclides is obtained as a weighted sum.

Limiting concentrations of nuclides in wastes have also been set according to the need for long-term isolation. The limits for ^{90}Sr and ^{137}Cs are the same as for high-level wastes, while the limit for ^{239}Pu is 0.1 Ci/m^3.

In the classification system used in Great Britain, five categories of wastes are defined on the basis of specific activity, form of radiation, and half-life of nuclides. Wastes that generate significant heat are grouped with high-level wastes in this system.[8]

France divides wastes into three categories:[8]

A. Wastes having a low or intermediate level of activity, containing nuclides with half-lives of less than 30 years, with a maximum activity of 3.7 GBq (0.1 Ci/t) for long-lived alpha-emitting nuclides.

B. Wastes having a low or intermediate level of activity, containing significant amounts of long-lived alpha-emitting nuclides.

C. High-level wastes containing long-lived nuclides and nuclides with high radiation levels.

A three-category system is also used in Italy; wastes are classified by the time required for the nuclides to decay to natural (background) levels.

A radioactive waste classification system proposed by IAEA in 1982-84 merits attention.[8,12,13] The system accounts for various disposal methods. Wastes are divided into five categories; two (with short-lived and long-lived nuclides) of intermediate-level wastes and two, similarly, of low-level wastes, plus one category of high-level wastes (Table 1). It is significant that the waste groups are not delimited by quantitative features as in, for example, the present NRB-76/87 classification.

Table 2, also proposed by IAEA, shows variants of the final disposal of radioactive wastes by category. Wastes belonging to categories IV and V (intermediate-level and low-level wastes containing short-lived nuclides) may be buried in liquid form (injected) in deep, permeable formations, or hardening slurries may be placed in relatively impermeable rocks. The use of nuclide half-life as the main classifying feature in analyzing the underground disposal of radioactive wastes is entirely justified because the requirements on disposal technology, geological formation, depth and location are largely dependent on how long the wastes remain toxic.

With regard to the practice of liquid radioactive waste disposal at Russia's nuclear industry sites, high-level wastes may also be split into two classes, those containing short-lived and long-lived

Table 1. A Classification of Radioactive Wastes Oriented Toward Final Disposal Concepts

Category of Wastes	Characteristics	Types of Wastes Within the Category
I. High-level wastes with long-lived radionuclides	High beta and gamma activity, significant alpha radiation, high radiotoxicity, high heat output.	High-level wastes from reprocessing of irradiated fuel and unreprocessed irradiated fuel in once-through fuel cycles.
II. Intermediate-level wastes with long-lived radionuclides	Intermediate beta and gamma activity, significant alpha radiation, intermediate radiotoxicity, low heat output.	Alpha wastes formed in reprocessing of irradiated fuel, fabrication of mixed oxide fuel, and operation of military enterprises engaged in producing and using plutonium
III. Low-level wastes with long-lived radionuclides	Low beta and gamma activity, significant alpha radiation, low to intermediate radiotoxicity, heat output insignificant.[a]	Wastes from operation, maintenance and decommissioning of nuclear power plants; wastes formed in front end of nuclear fuel cycle (refining, conversion, fuel fabrication) or at facilities producing and using radioisotopes in medicine, scientific research and education, industrial processes, etc.
IV. Intermediate-level wastes with short-lived radionuclides	Intermediate beta and gamma activity, alpha radiation insignificant,[a] intermediate radiotoxicity.	
V. Low-level wastes with short-lived radionuclides	Low beta and gamma activity, alpha radiation insignificant,[a] low radiotoxicity, heat output insignificant.[a]	

(a) This characteristic may be ignored in the decision how to dispose of the waste category.

nuclides. This distinction is based on the fact that the customary classification under the applicable standards[6, 10] fails to accurately reflect the possible hazards of particular wastes, especially in view of their location in a geological formation over extended periods of time. For example, high-level wastes having an activity of greater than 1 Ci/L and containing short-lived nuclides such as tritium, ruthenium, and ^{144}Ce, with half-lives of 1 to 2 years, present a different potential hazard and call for different handling from wastes having the same activity but containing long-lived transuranic nuclides with half-lives of hundreds or thousands of years (or more), such as isotopes of plutonium, americium, curium, and neptunium. While the first type of high-level wastes may pass into the class of intermediate-level or even low-level wastes after

Table 2. Options for Ultimate Disposal of Radioactive Wastes

Option	Environment	Category of Waste (see Table 1.1)				
		I	II	III	IV	V
1. Placement in deep geological formations (Note 1)	Dry (Note 2)	Solidified, packaged; placed with allowance for heat generated by radioactive decay	Solidified, packaged		Feasible, but such a high degree of isolation is not required for these categories of waste (Note 3)	
	Wet (Note 4)	Solidified, packaged; placed with allowance for heat generated by radioactive decay	Solidified, packaged		Solidified, may be packaged	
2. Placement in shafts or caverns (Note 5)	Dry	Not recommended	Possible under some circumstances	Possible with use of additional technical barriers	Solidified, packaged	Solidified, may be packaged
	Wet	Not recommended				
3. Placement in formations near the surface	Dry	Not recommended	Not recommended		Solidified, packaged	Solidified, may be packaged
	Wet	Not recommended	Not recommended		Solidified, packaged; with use of additional technical barriers	Solidified or packaged; with use of additional technical barriers
4. Injection of spontaneously hardening slurries into fissures formed in low-permeability beds of geological formations		Not recommended	Feasible with appropriate validation (demonstration) of technology and only for certain types of radionuclides		Applicable if appropriate technology is available	
5. Deep injection of liquid wastes into deep permeable formations		Not recommended	Feasible with appropriate validation (demonstration) of technology and only for certain types of radionuclides		Applicable if appropriate technology is available	

Notes:
1. Reservoirs specially constructed for ultimate disposal of radioactive wastes
2. Geological surroundings naturally isolated against penetration by groundwater.
3. Preferable for countries where near-surface formations present unfavorable geological conditions. Applicable if appropriate technology is available.
4. Geological surroundings where penetration of groundwater is possible.
5. Shafts or caverns formed naturally or by extraction of minerals or specially constructed for ultimate disposal of radioactive wastes.

The classifications of liquid radioactive wastes for deep injection take into account the origin of the wastes, the chemical and radiochemical composition, the frequency of delivery, and the use of pretreatment technology...

years or decades (depending on the initial activity), wastes of the second group will remain high-level wastes for an extended time.

In accordance with the standards and existing practice at Russian disposal sites, liquid radioactive wastes intended for deep injection are those that contain minimum amounts of long-lived TRU nuclides. The upper bound on the TRU content is dictated by safety concerns as well as the technical and economic possibilities of TRU recovery. Liquid wastes prepared for deep burial contain substantially less of these nuclides than the threshold value for long-lived TRU wastes in the U.S.

In some categories of liquid radioactive wastes destined for deep injection, the total content of uranium fission products having short and intermediate half-lives is below the limiting values for assignment to the high-level class under the U.S. system. At the same time, these categories of wastes must be classified as high-level under the standards now in effect in Russia.

The classifications of liquid radioactive wastes for deep injection take into account the origin of the wastes, the chemical and radio-chemical composition, the frequency of delivery, and the use of a pretreatment technology to ensure the wastes are compatible with the formation water and rocks. These wastes are divided into process and nonprocess solutions. Table 3 gives general characteristics of liquid radioactive wastes.[14, 15]

Process wastes are solutions generated in the actual reprocessing of irradiated fuel, including dissolution, after the uranium, plutonium, neptunium and americium have been extracted. Process wastes are classified as high level and intermediate level according to their radioactive nuclide content. Nonprocess wastes include solutions formed in auxiliary processes: effluents from special

Table 3. General Characteristics of Liquid Radioactive Wastes Received for Deep Injection Disposal

Category	Mode of Delivery	Total Specific Activity, Ci/L	Chemical Composition
Process wastes:			
Alkaline, intermediate-level	Continuous	10^{-6} to 1	Sodium nitrate, sodium aluminate, caustic soda
Acidic, arbitrarily classified as high-level	Periodic	0.1 to 10	Sodium nitrate, corrosion products, complexing reagents; pH = 1 to 3
Nonprocess wastes:			
Low-level wastes	Continuous	10^{-8} to 10^{-6}	Sodium nitrate, sodium sulfate, calcium carbonate, magnesium carbonate, detergents; pH = 8 to 10

laundries and equipment decontamination, leaks and trap waters, solutions produced in research work, regenerates from ion-exchange systems, liquors from the treatment of wastewaters, condensates, and so forth. Nonprocess solutions are usually low-level wastes, though some may be classified as intermediate level.

The radionuclide compositions of liquid radioactive wastes do not permit a clear separation by waste category. The nuclides present are isotopes of strontium, ruthenium, cesium, cerium, promethium and a number of other elements. It is possible to set apart the wastes that contain transplutonium elements: isotopes of americium, curium, californium and other elements, as well as neptunium (re-extract). Such wastes in liquid form are not candidates for deep injection disposal.

Wastes to be disposed of by deep injection are characterized by a prevalence of short-lived radionuclides. For example, nuclides with half-lives of about two years (isotopes of ruthenium, ^{89}Sr, ^{134}Cs, ^{144}Ce, ^{147}Pm, ^{95}Zr, and ^{95}Nb) account for around 50% of total activity in low- and intermediate-level wastes and up to 70% of total activity in high-level wastes. The balance of activity is supplied by nuclides with intermediate half-lives, chiefly ^{90}Sr and ^{137}Cs.

The preparation of liquid radioactive wastes for deep injection disposal usually includes the final extraction of long-lived nuclides, including uranium and isotopes of plutonium and the transplutonium elements. As a result, long-lived nuclides are present at unrecoverable levels in wastes for disposal, and they account for hundredths or thousandths of a percent of the activity.

Liquid radioactive wastes forwarded for deep injection are also classified as acidic (pH = 1 to 3) or alkaline (pH = 8 to 9) and as high-salt (up to 300 g/L) or low-salt solutions (1 to 10 g/L). Wastes having a dominant constituent are identified; examples are aluminate intermediate-level wastes formed by dissolution of the outer claddings of standard uranium blocks in irradiated fuel,[14, 15] manganese-niobium slurries, and others.

The frequency and size of liquid waste shipments arriving at the disposal site are further criteria for classification. Wastes may arrive periodically in batches of limited size, or be delivered in a virtually continuous manner over an extended period. For example, the experimental disposal of process wastes in an acidic medium (arbitrarily classified as high-level wastes) takes place periodically, in batches ranging from a few hundred to 2000 to 3000 m^3. Intermediate- and low-level wastes are disposed of in a practically continuous fashion or in steps lasting a few months with pauses for repairs and preventive maintenance. Before liquid radioactive wastes are sent to injection, they are pretreated; as a result, their characteristics may differ from those of direct process wastes.

1.2 Development of Technologies for Handling Liquid Radioactive Wastes

When the nuclear industry was established, the techniques used to handle radioactive wastes scarcely differed from those used in other industries.

When the nuclear industry was established in the late 1940s and the early 1950s, the techniques used to handle radioactive wastes scarcely differed from those used in other industries, particularly the metallurgical and chemical industries. The chief focus was on producing weapons-grade plutonium and uranium as quickly as possible and in the quantities needed. During those years, liquid wastes were discarded into existing surface waters or stored in special vessels. The possible dangers of radioactive wastes were obscured by the looming threat of a third World War and by the effort to achieve armed supremacy—or at least parity—in order to threaten a potential aggressor.

During this period of intense activity, the eldest member of the Soviet nuclear family, the Mayak Production Association in the Chelyabinsk oblast, released liquid radioactive wastes into the Techa River. From 1949 to 1956, 2.7 million Ci was discharged. These wastes then began to appear in existing waters, including the Karachai lake and marsh. High-level wastes were placed in special stainless steel storage tanks.[16]

The Siberian Chemical Combine, at Tomsk-7, built surface storage lagoons to hold intermediate-level wastes, a tank system for high-level wastes, and slurry basins for low-level wastes. In addition, natural waters were used for reactor cooling. A large treatment plant was operated at Area 13, where low-level wastes were pretreated by precipitation, filtration, ion-exchange treatment, and recycling of the treated water. Secondary wastes from the treatment facility—regenerates, mother liquors and slurries— were directed to surface reservoirs. Water that had been purified to potable standards (with respect to radionuclides) was recycled or released into streams.

Similar, though smaller, equipment was set up at the Mining and Chemical Combine at Krasnoyarsk-26. Here low-level wastes were pretreated and this portion of the operation later grew into a waste treatment plant. Intermediate- and high-level wastes were stored in special tanks. Contemporary nuclear plants in the U.S. used similar methods to handle their wastes, but instead of building exposed reservoirs for surface storage of liquid wastes, they tended to store these wastes in various types of reservoirs, both elevated and underground, or to discharge them into shallow soil layers through trenches, settling basins, and wells (cribs).[17]

In the Soviet Union, it soon became apparent that these means of discharging and storing wastes would have very unpleasant consequences. Radioactive contamination of the lower layers of the atmosphere in the immediate vicinity of plants such as

Mayak was 90% due to aerosol transport from the surfaces and banks of exposed waters and entrainment from storage vessels.[16] Pollution of streams, sometimes extending past the plant boundary limits, produced radiation burdens to the populace through external irradiation as well as ingestion of produce, fish, and water contaminated with nuclides.

In 1957, an explosion of nitrate-acetate salts in a high-level waste storage tank at the Mayak site released 20 million Ci of radionuclides that contaminated 20,000 km^2. Later, 600 Ci was entrained by winds from the shore of Lake Karachai and contaminated 1800 km^2.[16] These events proved the necessity of special handling for radioactive wastes and pushed the national government and the nuclear industry to take prompt action to prevent the harmful impact of wastes on people and the environment. Earlier, analyses of the nuclear industry had led many to believe that radioactive wastes should be treated to reduce the waste volume and to recover uranium, plutonium and other transuranic elements. Following would be conversion of the wastes to a stable, solid form to be stored and buried. Though research into this strategy was begun, these efforts were discontinued when practical problems connected with liquid radioactive wastes at major radiochemical plants became urgent. Developmental work had shown that a great deal of money—and, more importantly, time—would be required to devise technologies to fully process the wastes, including building facilities and designing and fabricating equipment. At that time, environmental contamination and the impact of those wastes on the public were allowed to continue.

An analysis of the ecological situation and the risk of accidents showed that the region of the Siberian Chemical Combine, barely 25 km from the city of Tomsk, stood under serious threat. Accidents similar to those in the Urals would lead to severe consequences for the people of one of Siberia's largest cities. The wastes presented a major hazard in case of war and air raid. At the suggestion of the Soviet government, a search for alternative technologies was instituted. One proposal was to place (inject) liquid radioactive wastes in deep reservoir formations similar to those in which crude oil and natural gas are found.

As the later chapters of this book will show, this approach proved correct and made possible controlling the negative impact of liquid radioactive wastes on personnel and the general public in the areas near the Siberian Chemical Combine at Tomsk-7, the Mining and Chemical Combine at Krasnoyarsk-26, and the Scientific Research Institute of Nuclear Reactors at Dimitrovgrad. The decision to establish deep injection sites for disposal of liquid radioactive wastes, and the early period of successful operation of these sites, did not detract from research toward other methods of waste handling, especially since unfavorable geology barred the largest nuclear enterprise, Mayak, from using deep injection technology.

In 1957, an explosion released 20 million Ci of radionuclides that contaminated 20,000 Km^2. Later, 600 Ci was entrained by winds ... proving the necessity of special handling for radioactive wastes.

While studies of processing and solidification technologies were vigorously pursued, practical results did not come until much later. The situation in other countries was similar.

Current views concerning the problem of how best to dispose of industrial and radioactive wastes can be summarized as follows:

- Any disposal or pretreatment technology chosen must produce the smallest volume of wastes and leave those wastes in the most stable form possible.

- The wastes that are generated, and their constituents, must be maximally used for any appropriate purposes.

- Wastes that cannot be used are to be concentrated, pretreated to render them minimally toxic and maximally stable, and stored until their use becomes possible.

- If wastes present a significant hazard to human beings and the environment when accumulated, pretreated, used, or stored, they must be placed in special, inaccessible, protected facilities or buried in geological formations after appropriate pretreatment. The use of the most advanced technologies for treating liquid wastes does not rule out using deep injection for some categories of such wastes.

Environmental protection and waste handling sometimes cost far more than the market can bear. Compromise has been the solution in most cases...

Frequently, the above principles cannot be completely implemented because of the conflict between the imperative to derive a salable product for maximum profit and the necessary and substantial expense of overcoming the harmful impact of wastes on the environment. Environmental protection and waste handling sometimes cost far more than the market can bear.

Compromise has been the solution adopted in most cases to ensure the profitability of industrial production and afford sufficient protection to human beings and the environment. In this way, methods of handling wastes have become acceptable even though the wastes have not been rendered completely harmless. The greatest compromises come about when vitally important operations must be maintained, for example, in the defense, food, energy, or transportation industries. In such cases, it is essential to find the optimal combination of waste handling methods so that even if the requirements for total elimination of harmful factors are not entirely satisfied, the consequences of these factors are held below some threshold at which irreversible effects are seen.

The choice of a method, or combination of methods, for handling liquid radioactive wastes is dictated not only by the chemical and physical characteristics of the substances, but also by the volumes that must be processed. At small research centers and plants

generating up to 10 to 20 m³/d of low- and intermediate-level wastes, processing and using the wastes presents no real difficulty. At large industrial plants where the volume of liquid radioactive wastes may range up to hundreds or thousands of cubic meters each day, the creation of waste-processing systems becomes a complex, sometimes insoluble, problem. For high-level wastes, processing and solidification become difficult at volumes of just a few cubic meters a day.

Technologies that have been developed and are in use for pretreating liquid radioactive wastes do not render the wastes totally harmless, but simply reduce their volume and convert them to a more stable form (e.g., immobilization in glass, concrete, or bitumen). These technologies generate "secondary" wastes, which require special handling. The operation of such systems calls for thorough measures to protect human beings against radiation. The traditional volume reduction methods for low- and intermediate-level liquid radioactive wastes are chemical precipitation, ion exchange, evaporation, filtration, diaphragm methods, immobilization in bitumen, and vitrification.

When precipitation is used, radioactivity is removed from the wastes not only through precipitation itself, but also through coprecipitation and adsorption of nuclides on the resulting bulky sediments in the settling system, and through physical capture of suspended colloidal particles by the sediments. A variety of chemical reagents are used: hydroxides of iron, aluminum, and titanium, phosphates, sulfates and sulfides; and ferrocyanides of copper, zinc and nickel. The process yields a liquid and a solid phase. The purification of the liquid phase is characterized by ratios of 50 to 100 or higher. The solid phase is enriched in the nuclides. The liquid phase may be subjected to further treatment and then recycled or discharged to the environment. The solid phase must be processed and then stored or buried. Precipitation is used chiefly for low- and intermediate-level wastes.

The treatment of liquid radioactive wastes by ion exchange involves natural and synthetic inorganic and organic materials. The natural inorganic materials include clays, zeolites, and minerals such as vermiculite and clinoptilolite. Organic ion-exchange materials are primarily polystyrene and phenol-formaldehyde resins including functional groups. While ion-exchange resins are very effective in removing radionuclides from water (10^2 to 10^4 ratio), they impose strict requirements on the feed streams, which must contain 1 g/L of salts and less than 4 mg/L of suspended solids. As a result, the effluents must pass through a pretreatment step. The secondary wastes of ion-exchange systems are liquors from the washing of the resins (regeneration solutions), containing substantial amounts of nuclides and salts, and spent ion-exchange materials, which must be further processed, stored, or buried.

Evaporation is widely used for processing wastes and gives a 10^4 purification ratio (10^6 in some systems). The wastes must be pretreated to prevent corrosion of the evaporators, frothing, and scaling. The secondary wastes comprise bottoms and slurries with high nuclide and salt contents. Some part of the nuclides may remain in the condensate, so that evaporation must be done in several stages. This process is very energy intensive and becomes less efficient when used on large volumes of wastes.

Filtration is usually an adjunct process in preparing liquid radioactive wastes for the other pretreatment methods. A variety of filter, centrifuge, and wet cyclone systems are used. The filter media and solids must be further treated.

Diaphragm processes include reverse osmosis, electrodialysis and ultrafiltration; these are applied mainly to low-level wastes. Currently under development are electro-osmosis, and electrochemical ion exchange. All these processes yield secondary wastes that require special handling.

The final stage in processing wastes for disposal is solidification of the wastes, chiefly the high-level wastes that pose the greatest hazard. The most advanced method is immobilization of high-level wastes in glass; borosilicate, phosphate, basaltic, soda-lime and other types of glass are used. In some processes, liquid wastes are dehydrated and calcined before solidification. Vitrification processes are carried out at high temperatures and entail gas and aerosol discharges, which call for special precautions.

High- and intermediate-level wastes can be incorporated in special cement mixes based on portland cement. The development of mineral-like, ceramic, and metalloceramic composites is now in the research stage. Immobilization in bitumens is practiced on low- and intermediate-level wastes; the combustibility of bitumens is a negative characteristic of this technology. Another method of direct radioactive waste disposal under research and development is transmutation of radionuclides in nuclear reactors and similar physical facilities. Long-lived nuclides are transformed to short-lived ones, which decay in comparatively short times.

Each waste processing method described here is quite complicated, requires special apparatus with anticorrosion features, consumes a large amount of energy, and presents certain hazards to operating personnel. Accidents and contamination of the surroundings are always possibilities when such process systems are in operation.[18] Each process generates secondary wastes or concentrates that must be stored and buried. The deep injection of liquid radioactive wastes immediately after they are formed offers a remedy for some of these difficulties and involves much lower costs. From the 1950s into the 1980s, deep injection disposal was the sole feasible technology for some kinds of wastes.

1.3 Basic Safety Requirements for Handling Liquid Radioactive Wastes

Safety requirements for the handling of liquid radioactive wastes are based on standards developed by the United Nations Scientific Committee on Effects of Atomic Radiation (UNSCEAR) and the International Commission on Radiological Protection (ICRP), the IAEA Recommendations, and national standards (in Russia, the Radiation Protection Standards NRB-76/87).[6, 19, 20, 21]

The main objective of the safety requirements is to afford protection from ionizing radiation to individual persons, their posterity, and the human race as a whole, and at the same time to create appropriate conditions for necessary practical activity of human beings during which they may be subjected to the impact of ionizing radiations. The ICRP and other international organizations recognize the linear, nonthreshold hypothesis of the effect of radiation, which can be summed with the effects on humans due to chemical pollutants in the environment. For this reason, limiting doses are established to guarantee a certain degree of safety, and it is recommended that radiation burdens on the population as well as plant workers be reduced to minimal levels.

The fundamental principles applied to radiation burden in the ICRP recommendations can be stated as follows:[19]

- Ionizing radiations may be used only in cases of real and clear benefit.

- All received doses must be held as low as can reasonably be achieved in view of economic and social factors.

- The equivalent received dose of individual persons must not exceed a limit recommended in the standards for corresponding conditions.

The NRB-76/87 standards now in effect also embody the following principles:

- Nonexceedance of an established basic dose limit.

- Avoidance of any unjustified exposure.

- Reduction of radiation dose to the lowest possible level.

Compliance with these requirements as applied to the handling of radioactive wastes has some special features. Wastes are substances and materials considered unsuitable for further use. Accordingly, the trend toward disposing of wastes as soon as possible after their formation, with minimal negative impact on human beings and at minimal expense, is fully justified. A radical solution to

The main objective of the safety requirements is to afford protection from ionizing radiation to individual persons, their posterity, and the human race as a whole.

the problem is to use rocketry and space technology to dispatch radioactive wastes beyond the bounds of the solar system, or toward the sun, or to convert their components into nonradioactive substances by transmutation. At present, however, realization of these technologies on an industrial scale is far off.

A more practical means of handling radioactive wastes is to localize and isolate them in special facilities or in appropriate natural objects so that the wastes cannot act on human beings. Keep in mind that the intermediate operations on the wastes (conditioning, processing, purification, etc.) entail exposure of plant workers and, to some extent, the public as a result of ventilation releases. The level of irradiation can be held below established limits, but the nonthreshold concept makes such exposures undesirable. The deep injection disposal of liquid radioactive wastes, usually carried out just after formation of the wastes and involving a minimum of preparatory operations, has certain advantages from this standpoint.

Any analysis of radioactive waste burial must take account of the possible long-term consequences, in particular the relationship of later generations to the disposal sites and any extra work necessitated by the wastes in the future. The IAEA, *Safety Principles and Technical Criteria for the Underground Disposal of High-Level Radioactive Wastes*[21] includes the following recommendations: Future generations of people must bear the minimal burden of concern for buried high-level wastes in view of technical, social and economic factors. The safety of a high-level waste storage site after shutdown and after the passage of the specified period of time must not be based on active monitoring and maintenance of the site. The degree of isolation of high-level wastes must be such that the future risk to humans is no greater than the present risk. Attention is drawn to the necessity of preventing cross-border effects when a waste disposal site exerts an influence on contiguous countries or when it can pass to another country upon a change of frontiers. The requirement of minimum possible irradiation of human beings and nonexceedance of dose and risk limits is affirmed. A number of technical criteria are set forth, including the requirement that high-level wastes be converted to solid form before burial.

As will be discussed in this book, the deep injection of liquid radioactive wastes satisfies these requirements to some degree. The exception is the condition of preconversion of high-level wastes to solid form. As follows from the radioactive waste classifications of Section 1.1 and the material presented in Chapter 5, however, high-level wastes that can be disposed of in solution form occupy—in respect of some characteristics—a position in between high-level and intermediate-level wastes and are actually closer to the latter.

Thus the main objective of the various methods of handling radio-active wastes is to prevent all impacts of the wastes on human beings. An analysis of the current standards, methodological recommendations, and scientific publications makes it possible to formulate the following requirements on the final stages of radioactive waste handling:

- The wastes must be isolated from the surroundings in which human beings live and directly act and in which animals and plants live and grow.

- The waste storage or disposal site must not be vulnerable to accidental or deliberate penetration. The wastes must not be subjected to the action of natural catastrophes that could extract them from the geological formations.

- The boundaries of the facilities, territory or geological medium in which the wastes are located must be clearly defined and established with allowance for possible natural phenomena. Activity not connected with the wastes is forbidden or restricted within the boundaries of the storage or disposal site.

- Isolation of the wastes within the established boundaries must be ensured for the requisite time until the nuclides or other components cease to pose a hazard to human beings and the environment or until the currently forecast period of time has passed.

- To diminish the risk of irradiation of personnel and the general public, preliminary operations on wastes (preparation, processing, transportation) that are accompanied by the release of radioactivity into the surroundings or by the impact of radiations must be minimized.

- When radioactive wastes are stored or buried, the volume of the storage site must not be host to processes that degrade the conditions of waste isolation, lead to the escape of waste constituents beyond the boundaries of the storage site, or necessitate special operations for the continued storage or reburial of the wastes.

- Radioactive waste storage or burial sites must occupy the least possible area and volume, exert the minimum effect on natural resources and various forms of activity when used in adjacent territories.

Specific requirements on radioactive waste handling should be based on the composition and properties of the wastes in question; on existing scientific, technical, social, economic, public health, and ecological factors; and on historical traditions. For example, such requirements may include conditions for storage and burial of certain types of waste in solid or liquid form, quantitative characteristics, and equipment design. These requirements can be included in national standards and technical codes developed for specific periods of time and can allow for technical and economic feasibility and other considerations.

Information on the standards and technical regulations that form the basis for deep injection methods used with liquid radioactive wastes is presented in Chapter 4.

2 Geological Requirements for Deep Injection Disposal of Liquid Radioactive Wastes

The disposal of liquid radioactive wastes in deep geological formations, like the disposal of other harmful substances, and objects that can harm human beings, is a logical extension of society's millennia of efforts to use the Earth's natural resources to satisfy humanity's daily needs. A most vital requirement is protection against the harmful consequences of society's own activities.

One of the Earth's richest natural resources is the subsurface, the traditional source of minerals. The range and volume of minerals that have been mined, as well as the depth to which humans have gone in search of them, have increased continuously. By the early 1990s, 4 billion tons of coal, 900 million tons of iron ores, 2 billion tons of crude oil, and 2 trillion m^3 of natural gas were being extracted from the Earth each year. The planet's subsurface is also put to heavy use as a source for water supply systems and for producing mineral waters. A variety of wastes, including radioactive wastes, arise from the treatment and use of these resources to obtain needed substances, materials, and energy. The return of these wastes underground, together with other methods of handling them, is quite natural.

At the present level of industrial development, the value of the subsurface is increasing, both as a source of mineral raw materials and as a medium for the construction of underground facilities (i.e., manufacturing complexes, transportation, utilities, storage reservoirs [e.g., wastes], and underground nuclear power plants). Intensive use of the Earth's interior for extraction of minerals, research in connection with prospecting, exploration and development of deposits, and studies of the planet's crust have yielded a huge body of knowledge about the structure and properties of the geological medium and the laws describing it.

Based on these data, a scientific-methodological basis has been established for dealing with critical problems in using the Earth's resources. Analysis of these geological data, and practical experience in mining and in crude oil production gives reason to conclude that it is possible and feasible to place (bury) liquid radioactive wastes in deep strata of the Earth's crust that meet certain requirements. This conclusion is based first of all on the following facts and relationships:

- The isolation of deep aquifers and complexes from the surface; the vertical hydrodynamic and hydrogeochemical zonation

...the value of the subsurface is increasing, both as a source of mineral raw materials and as a medium for construction of underground facilities.

of subterranean waters; the low rate of mass transport in subsurface waters.

- The widespread "hollowness" of rocks, represented by intercommunicating pores and fissures, which enables rocks to contain water, petroleum, and gas as well as various solutions injected under pressure.

- The formation and preservation of oil and gas pools and deposits of minerals, including uranium, through the retention in rocks of chemical compounds and radioactive substances from waters flowing through the pore space.

- The stability of the Earth's crust; the difficulty of gaining access to deep formations either accidentally or deliberately.

At the same time, opponents of using the geological medium for the disposal of liquid radioactive wastes, and industrial wastes in general, express different views based on the following facts:

- The crust, and particularly the water-saturated zones of the crust, is permeable to some degree, in part because of deep tectonic faulting. There are well known karst regions where the rates and volumes of movement of waters are quite high.

- Natural changes in the Earth's crust are accompanied by earthquakes and may affect the conditions of isolation of deep geological formations.

- Subterranean waters are used for sanitary water supplies, and also as the source of curative waters and industrial process water. The deep injection of liquid radioactive wastes and industrial wastes diminishes the natural reserves of subsurface waters that could be used in the future.

Clearly these conflicts can be resolved through a well thought-out and scientifically justified policy of resource management. Such policy must provide guidance for the following: selection and preliminary examination of geological formations at sites under consideration for waste disposal; analysis of the compatibility of the wastes with the rocks and subterranean waters; forecasting of burial processes and changes in the geological surroundings; and sound engineering decisions. With an eye toward the concerns of opponents, the remainder of this chapter examines the conditions of geological formations that would be suitable for deep injection disposal of liquid radioactive wastes.

2.1 Vertical Zonation of Subsurface Waters

The practice of underground disposal of liquid or solid radioactive wastes must take into account the structure and properties of the upper portions of the Earth's crust. The zone of concern comprises the lithosphere and the subsurface part of the hydrosphere, which is located inside the lithosphere and includes all the Earth's waters beneath the exposed surface and the floors of surface waters. Subterranean waters in both unconfined and confined conditions can to some degree come into contact with buried radioactive wastes, alter them, and cause waste constituents to migrate away from the disposal site.

Subsurface waters play a major part when liquid radioactive wastes are buried in reservoir formations. The hydrodynamic and hydrogeochemical conditions determine whether burial is possible and safe, govern the size of the area of influence of the burial, and thus govern how much space is made unavailable for mining. Subsurface waters also influence what technology must be used to prepare wastes for burial, and how the disposal site must be designed. The burial (injection) of liquid radioactive wastes inevitably brings with it an interaction of the wastes with subsurface waters, because the wastes are placed in natural voids, that is, in the pore space of the reservoir rocks, which contains formation water.

The volume of water in the subsurface part of the hydrosphere is comparatively small, only about 0.63% of the total water in the Earth's hydrosphere, estimated at 1.36 billion km^3. Most water is contained in the oceans (97.2%); the polar icecaps and glaciers account for 2.15%; fresh and salt lakes and inland seas, 0.017%; the atmosphere, 0.001%; vadose waters (including soil moisture), 0.005%; and river channels, 0.001%.[22] World reserves of fresh water come to a little over 30 million km^3, about 97% being concentrated in the polar icecaps and glaciers; about 2% in the atmosphere, freshwater lakes, rivers, and soil moisture; and roughly 1% in subsurface fresh waters. Mineralized waters make up 70 to 80% of the total water volume of the subsurface hydrosphere.

The distribution of water in the hydrosphere, its composition, physical and chemical properties, its rate of movement, and the rate of transport between the atmosphere and the surface and subsurface parts of the hydrosphere are strongly dependent on which layer of the Earth's structure is host to the water and on the structure and properties of the natural formations enclosing the water. The time required for the transport of water between the atmosphere and surface waters may be from hours to days, and water masses in the atmosphere and in rivers can move over great distances in short spans of time. In contrast, the rates of natural movement of subsurface waters in deep aquifers containing highly mineralized waters (brines) usually amount to centimeters or meters per year. The movement is not uniform in direction, and the direction of flow may change frequently over geological epochs. There is virtually no exchange of water between such aquifers and surface or shallow subsurface waters. The basic idea in deep injection disposal of liquid radioactive wastes is to use portions of the geological medium where natural conditions ensure localization of the wastes at the place of burial, despite the presence of subsurface (primarily saline) waters whose natural movement is insignificant.

A typical feature of the subsurface hydrosphere is hydrodynamic and hydrogeochemical zonation, expressed in the pressure variation (hydrostatic and geostatic compression) of subsurface waters, their chemical makeup, the composition of dissolved gases, and other characteristics. The vertical zonation of subsurface waters is very marked in platform regions (Russian Platform, West Siberian Plate, etc.). Zonation reflects the conditions of formation and the structure of the kilometers-thick sedimentary strata where the geologic section is horizontal or monoclinal and beds of permeable rocks (sands, sandstones, limestones) alternate with beds of less permeable rocks (clays, argillites, salts) capable of blocking the movement of water.

According to the classification used in hydrogeology, the following component elements are distinguished in strata of water-saturated sedimentary rocks:

A *water-bearing bed* is a bed of permeable rocks uniform in lithofacial makeup, filtration (flow) properties and holding capacity. Voids (pores or fissures) in the rock contain water, which can move by gravity or along pressure gradients.

A *water-bearing horizon* or *aquifer* is one or more water-bearing beds with common conditions of formation and movement of subsurface waters. The aquifer is hydrodynamically self-contained, as manifested in the close interrelationship of the head in all beds.

A *water-bearing complex* or *aquifer system* comprises several aquifers separated by relatively impermeable or "tight" rocks and having common resources and composition of subsurface waters. The horizons making up the system may be hydrodynamically connected, although they may differ in head.

A *confining bed* (horizon) is made of rocks that retard but do not prevent the flow of water; the water in these rocks is chiefly in the bound condition, and the pores are so small that capillary forces hinder its movement. The confining properties of clay rocks or their ability to block the flow of a liquid depends on the specific conditions, including the temperature and the salt content of the pore fluid. Beds of salt, gypsum, and anhydrites fall into the class of confining beds.

A *confining system* consists of several confining beds or horizons interbedded with thin water-bearing beds. The system is usually a regional feature isolating overlying or underlying aquifers or aquifer systems.

Confining systems in the geologic section are a necessary condition for the deep injection disposal of liquid radioactive wastes. The bedding of the sediments in the complex and the presence of rocks differing in plasticity, including permeable rocks, ensure that the system will retain its confining properties when the hydrostatic pressure changes as a result of injecting liquid wastes in underlying aquiferous horizons.

The following terms are used to describe the movement of subsurface waters in aquifers and aquifer systems:

- The *effective flow velocity* is the average (resultant) rate of motion of water molecules in the pores, in the direction of flow projected onto a plane approximately parallel to the "roof" or "floor" of the aquifer (i.e., its upper or lower boundary).

- The *residence time of water* is the time required to replace the full initial volume of water in the horizon pore space by an equal volume of incoming water.

- The *age of subsurface waters* is the length of time from the water's entrance into the horizon until the present.

The waters of deep horizons that are suitable for the disposal of liquid radioactive wastes are under pressure. In wells that penetrate the horizon, the water level is significantly higher than the top of the aquifer. This pressure head is due to hydrostatic and geostatic pressure, the flow resistance of the porous rocks to subsurface waters moving within, and the development of tectonic stresses in the Earth's crust. Naturally, the creation of the head regime requires the presence of relatively impermeable beds and systems in the section that can block the vertical redistribution of the subsurface waters.

Fresh waters occurring in the first tens of meters beneath the surface that are closely connected with the surface waters take on a pressurized character with increasing depth and are replaced by slightly saline waters having a hindered connection with the surface waters. Saline waters (brines) are farther down, usually at depths of more than 1000 m. The anion composition of the waters in the upper part of the section is dominated by hydrocarbonates; sulfates begin to appear at greater depths, and chlorides farther down. The cation composition shows a similar progression, though less marked: calcium and magnesium dominate in the upper horizons, sodium and calcium in deeper ones. "Trace" components exhibit regular variations with depth. The contents of iron, manganese, silica, barium, and many metals decline with depth, while the contents of bromine, potassium, strontium, and radium increase with depth. The redox potential decreases with depth; oxygen disappears; nitrogen, methane, hydrogen sulfide, carbon dioxide, and heavy hydrocarbons build up. Oil and gas pools may, under certain conditions, survive for millions of years in deep aquifer systems.

Based on studies of the regime and composition of subsurface waters in thick sedimentary strata, and on experience gained in exploring and developing oil and gas fields and deposits of fresh and iodide-bromide waters, geologists have identified "stacked" hydrodynamic and gas-hydrochemical zones. These zones differ depending on the regime, genesis, and movement of the subsurface waters; their composition; and the nature of the energy potential involved in generating their head. (See Figure 1.)

A variety of subsurface hydrologic zone classifications, based on several different criteria, have been proposed by N. K. Ignatovich, N. A. Gatal'skii, Yu. V. Mukhin, G. Yu. Valukonis, G. V. Bogomolov, N. K. Zaitsev, and others.[23] In general terms, they define a zone of free (active) water exchange or movement, a zone of hindered water movement and a stagnation zone (greatly hindered water movement).

The *zone of free water movement* occupies the upper portions of the section, to depths of a few hundred meters (zone of aeration, groundwaters, unpressurized interlayer waters, pressurized formation waters). The energy potential is hydrostatic. The waters originated as precipitation entering the system by infiltration in the recharge area, and then moved along the beds and aquifers toward the discharge area. The head (hydrostatic pressure) in

Name	Typical Depths	Characteristics and Genesis of Subsurface Waters	Nature of Energetic Potential Movement of Waters	Time of Complete Water Exchange (residence time), years [a]	Typical Velocities of Water Movement	Possibility of Industrial Waste Disposal
Zone of free (active) water movement	Few hundred to 1000 meters	Free waters fresh; mineralized lower	Hydrostatic; from recharge to discharge	10s to 1000	Meters to 10s of meters per year	Exceptionally in lower regions
Zone of hindered (impeded) water	Few hundred to 1500-2000 m	Free waters ancient, mineralized in upper section; connate and elisional at greater depths	Hydrostatic; from recharge areas to injected portions of basin	1000s and 10000s of years	Fractions of a meter to a few meters per year	Possible when conditions are favorable
Zone of retarded water movement (stagnant regime)	From 1500-2000 m to 3000-5000 m and deeper	Highly mineralized; chiefly connate and elisional	Tectonohydraulic and genostatic; governed by development of tectonic stresses and characterized by change in direction over time	Millions of yers	Centimeters per year	Possible when conditions are favorable

Figure 1. Vertical Zonation of Subsurface Waters

pressurized horizons results from the difference in elevation between the recharge and discharge areas, the resistance of the rocks to water flow, and the geostatic (overburden) pressure. From the industrial waste disposal standpoint, the lower parts of this zone are of interest. The flow velocity is no more than a few meters a year, while the residence time and the age of the subsurface waters are estimated in thousands and tens of thousands of years.

The *zone of impeded water movement* lies between the zone of free water movement and the stagnant regime; in most cases it is found at depths of a few hundred to 1000 to 1500 m. Waters in this zone move from recharge areas toward submerged parts of the basin at velocities from centimeters to a few meters per year. The energy potential is hydrostatic. Along with infiltration in the upper part of the zone, sedimentation processes formed the subsurface waters (buried "trapped" waters). These waters, which were present in the pore space of the rocks that formed in sedimentary basins (ancient seas and lakes), have gone through the stages involved in lithogenesis along with the rocks. The lower parts of the zone contain "elisional" waters, which have been squeezed out of overlying and underlying plastic rocks by the geostatic pressure (i.e., the weight of the strata). The age of the subsurface waters is slightly younger than the age of the rocks, ranging from hundreds of thousands to tens of millions of years. Water in this zone moves primarily between the lower horizons of the free water movement zone and the horizons of the impeded water movement zone as well as within the zone. Permeable horizons existing throughout this zone can be used for deep injection disposal of liquid wastes.

The *stagnant zone* occupies the lower portions of the sedimentary cover at depths from 1500 to 2000 m and greater. The energy potential here is geostatic and tectonic. The heads of the subsurface waters are largely determined by the geostatic pressure and the tectonic stresses. Changes in the tectonic stresses bring about a tectonohydraulic movement of subsurface waters. The velocity is measured in centimeters per year, and the direction of movement can change repeatedly over the geologic epochs. The waters contain large amounts of salts and are usually brines of sedimentary origin. The fraction of elisional waters and waters formed on hydration of clay minerals increases in the lower portions of the zone. These phenomena lead to the formation of inversions, where waters with low salt contents lie below brines, being separated from them by relatively impermeable beds. The existence of inversion zones, where waters lower in salt content and hence lower in density occur beneath denser waters, demonstrates the good isolating qualities of the relatively impermeable beds in deep horizons.

The age of the subsurface waters in the stagnant zone is close to that of the enclosing rocks: tens, hundreds, or thousands of millions of years. The main water transport process is between

*Preliminary
studies make
it possible
to determine
the degree
of isolation
of aquifer
systems, to
identify the
interrelations
between
horizons,
and to avoid
creating
deep injection
disposal sites
where conditions
are unfavorable.*

aquifers and aquifer systems within the zone, and also with waters in fissured zones of the underlying foundation (crystalline basement) as a result of recent tectonic movements and the processes of subsidence, uplifting, compression, and tension. The stagnant zone has practically no connection to surface waters and shallow groundwaters. The zone of impeded water movement may play some part in the formation of the waters of this zone.

The permeability of the aquifer rocks in the stagnant zone decreases with depth, while the isolating ability of the clay rocks becomes less through loss of plasticity under the increasing pressures and temperatures (lithification) as well as through jointing. For these reasons, the upper and middle portions of the zone are best suited for disposing of liquid wastes.

The zones of subsurface waters are usually separated by confining systems of regional extent, while confining beds inside each zone play an important role in creating the hydrodynamic and hydrogeochemical zonation. The criteria for demarcating the zones are the composition of the subsurface waters, the head regime, the depth, and the characteristics of the enclosing rocks. Important measures are the composition of dissolved gases (especially the presence of helium) and the age of the subsurface waters as determined from the content and ratio of various isotopes such as tritium, ^{14}C, uranium, deuterium, helium, and argon.

At the same time, the zonation is known to break down, for example, where waters with the features of deep and ancient waters (elevated salt content, components typical of waters in the impeded movement and stagnant zones, high levels of helium, abnormal heads, anomalous temperatures) are encountered at comparatively shallow depths. These phenomena occur in the areas of deep tectonic fractures (faults), regions where there is a vertical permeable feature connecting the aquifers and aquifer systems at different depths, or strongly jointed areas resulting from tectonic causes. Such phenomena are commonly confined to marginal zones of fault blocks, geosutures, rifts, aulacogens, and other manifestations of deep tectonic structures of the Earth's crust.

Obviously the regions where these tectonic phenomena occur are unsuitable for the deep injection disposal of liquid radioactive wastes and industrial wastes. Preliminary studies make it possible to determine, with a fair degree of confidence, the degree of isolation of aquifers and aquifer systems, to identify the interrelations between horizons, and to avoid creating deep injection disposal sites where conditions are unfavorable. However, not all tectonic fractures found in platform regions and cutting through sedimentary strata, aquifer complexes, and aquifers in different hydrodynamic zones function as vertical connecting paths. In most cases, they are barriers offering no vertical connection.

The vertical zonation of subsurface waters is widespread in Russia, that is, virtually everywhere in the Russian Platform, Caspian Basin, Urals Foredeep, and West Siberian Platform. Regions of vertical zonation are in good agreement with regions favorable for the deep injection disposal of industrial wastewaters, according to the 1:2,500,000 map prepared by the Ministry of Geology in 1969 under the direction of Academician A.V. Sidorenko.

2.2 Reservoir Properties and Isolating Qualities of Rocks

Most rocks have natural cavities; that is, intercommunicating pores or fissures of various sizes are present. As a consequence, porous rocks can enclose water and various solutions. Under natural conditions, the pores in permeable rocks are filled with subsurface waters that can move in the pores when a pressure gradient is imposed (Figure 2).

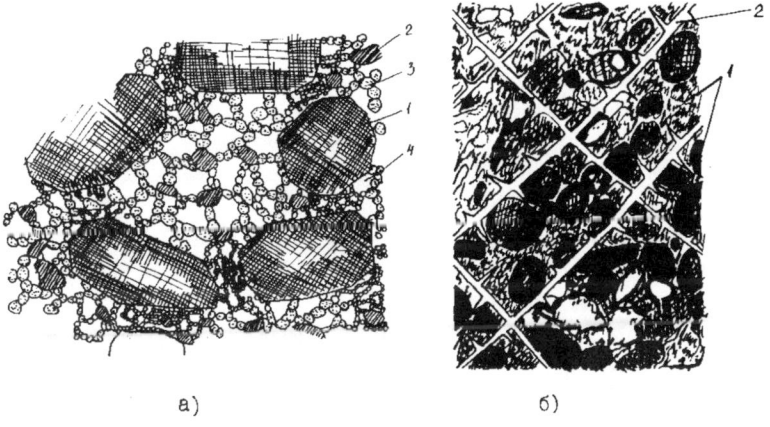

a)   б)

Figure 2. Structure of Sedimentary Rocks. (a) Rock containing sandy and clayey constituents: 1 - sand grains, 2 - fine powder grains, 3 - clay particles, 4 - pores. (b) Carbonate rock (an organic limestone formed from ooze): 1 - blocks, 2 - fractures.

The porosity of a rock is measured by the ratio of the aggregate pore volume to the total volume of the rock; the ratio is stated either as a percentage or as a fraction. Permeable rocks making up beds suitable for the disposal of liquid wastes have typical porosities of 5 to 45%. In practice, an effective porosity is used; this relates to the pores that communicate among themselves and can be filled by the injection of solutions. The remaining pores are dead-end or subcapillary pores.

Two types of porosity are distinguished: primary and secondary. Primary porosity is present in sediments when they are deposited, and after lithification, some porosity is retained. Primary porosity is

Most rocks have natural cavities ...pores or fissures are present. As a consequence, porous rocks can enclose water and various solutions.

characteristic of poorly consolidated rocks such as sands, weakly cemented sandstones, and some limestones. Such rocks contain the following kinds of pores or interstices: supercapillary, with diameter greater than 0.5 mm, in which liquid can move freely; capillary, with diameter 0.0002 to 0.5 mm, in which liquid can still move but only under certain conditions; and subcapillary, with diameter smaller than 0.0002 mm, in which the action of intermolecular forces is so strong that the liquid in the interstices cannot move under the observed pressure drops. The largest pores in sandy-clayey rocks are usually no larger than 1 mm. Secondary porosity occurs after portions of the rock are dissolved or fractured. A hybrid of primary and secondary porosity is often seen in consolidated sedimentary rocks where the interstices comprise fractures that divide the rock into porous blocks. The fracture dimensions are commonly larger than the pores in the blocks.

Porous sedimentary rocks have a significant internal pore area, as much as tens of square meters per cubic meter of rock. Pore area is crucial because the physical and chemical interactions between liquid radioactive wastes and the rocks occur on these internal surfaces. Through sorption phenomena, these interactions can retard the dispersal of radioactive waste components.

The reservoir properties of rocks are also governed by permeability. When wastes are injected into a reservoir formation, permeability would allow dispersal through the pore space of the formation to a fairly great distance from the well under the small pressure gradients that can be established.

The permeability of rock may be measured by the "filtration factor," defined as the amount of a liquid flowing in the porous medium through a unit area of the permeable bed under a unit pressure gradient. The "permeability coefficient" or "permeability" also takes account of the viscosity of the flowing liquid. Because liquid radioactive wastes are very similar in viscosity to subsurface waters, the filtration factor is commonly used in practice instead of the permeability coefficient.

The permeable rocks in reservoir horizons used for the disposal of liquid radioactive wastes have filtration factors ranging from 0.5 to 5 m/d. Values of up to tens and hundreds of meters per day are seen, for example, in karst rocks or gravels. The high velocities of waters in such rocks make them unsuitable for the deep injection disposal of liquid radioactive wastes and industrial wastes. Low filtration factors, under 0.1 m/d, are typical of clayey rocks. The injection of solutions into these rocks is difficult because high pressures are developed at low injection rates.

The reservoir properties also depend on the thickness of the formation or bed, which is defined by the area between its lower and upper boundaries. An "effective" formation thickness is defined as the sum of the thicknesses of the individual permeable beds with the ability to absorb solutions on their injection. Effective thicknesses of reservoir horizons suitable for waste disposal range from tens of meters to 100 m.

The product of the effective thickness of the reservoir formation and the mean filtration factor is called "transmissivity"; typical values are from a few tens to a hundred square meters per day. The product of effective thickness and mean effective porosity is described as the "specific capacity" of the reservoir formation; it characterizes the available volume of pore space in the formation (in cubic meters) per square meter of formation area. Measured in meters, the specific capacity usually rangeas from a fraction of a meter up to 10 m.

Reservoirs are natural formations and are, therefore, irregular in both plan and cross section. Existing investigation methods make it possible to establish the character and the scale of such irregularities, evaluate their effect on the distribution of wastes, and arrive at a sound decision on whether to proceed with waste disposal.

Another important property of reservoir rocks is "sorption capacity," which describes their ability to retain substances dissolved in the wastes—chiefly radioactive nuclides. Sorption processes take place in the pore space and cause radioactive nuclides and some chemical compounds to be converted to the solid phase and held up by the rocks. The sorption processes are partly irreversible; that is, only part of the nuclides are desorbed by the flow of subsurface waters, while the remainder go into the pore liquid over long spans of time through leaching and diffusion. Similar phenomena are seen under natural conditions in the formation of some mineral deposits. (The results of studies of the retention properties of rocks at liquid radioactive waste disposal sites are presented in Chapter 3.)

Liquid radioactive wastes in the pore space of a reservoir formation must be isolated from the surface, shallow groundwaters, and overlying aquifers. The barrier or "roof" of a reservoir horizon is formed by confining systems and confining beds, which are widespread in thick sedimentary strata and consist of beds of various clay rocks (clays, clay shales, marls, argillites), less frequently by salts and secondary beds of permeable rocks.

The isolating qualities of confining systems are manifested under natural conditions where these formations separate hydrodynamic zones as well as aquifers and aquifer systems differing in the pressure and chemistry of the subsurface waters. Confining beds and systems owe their isolating qualities chiefly to beds of clays. Clays are highly porous rocks, but their interstices are subcapillary in size, and the intermolecular forces in them prevent flow. The most important components of clay rocks are clay minerals (kaolinite, hydromica, montmorillonite, and chlorite) and fine detritus (pelite). The clay particles are usually smaller than 0.004 mm (pelite particles smaller than 0.001 mm). The clay rocks contain impurities, mainly aleurolitic, and to a lesser degree sand-like grains of quartz, feldspars, muscovite, and other minerals.

Special conditions must exist if flow processes are to take place in clays: the pressure gradient must be greater than the threshold values, the temperature must be raised, and the salt content of the contacting waters must increase. When flow occurs, the volume rate and flow velocity are small because clays have small filtration factors, usually in the range of 10^{-3} to 10^{-6} m/d.

It is typical of clays that their barrier and flow properties depend on a number of factors; this point must be considered in relation to the deep injection disposal of industrial wastes. If the pressure increases in the underlying or overlying horizon, the permeability may either decrease or increase.[24] In a clay reservoir, the replacement of the formation fluid by liquid radioactive wastes with low-salt content leads to a decline in permeability of the clay rocks overlying the reservoir; if the temperature rises, the permeability increases slightly.

Because these confining systems are composed of beds of plastic and permeable rocks, these systems remain continuous and retain their isolating qualities even when pressures are elevated by waste injection and when neotectonic and recent movements occur in the Earth's crust. As depth increases, the clays decrease in porosity and permeability, undergo changes in form, and are transformed to argillites and clay shales. At greater depths, the formation of joints may increase their permeability. Salt-bearing sediments and marls now begin to function as the roof. Losses of screening power by the clay rocks and resulting vertical leaks between underlying and overlying horizons are usually seen in tectonic fault zones where there is a permeable plane or where zones of sandy character appear in the clay rocks (lithologic "windows"). Such places can be reliably identified in preliminary studies; a disposal site is chosen so that waste injection does not affect the condition of overlying horizons.

2.3 Isolating Qualities of the Geologic Medium as Manifested in the Formation of Mineral Deposits

The deep injection disposal of liquid radioactive wastes is accompanied by several processes in which substances are redistributed between the solid and liquid phases, and waste components accumulate as nonuniformities in the geologic medium. Analogous processes take place under natural conditions in the formation of mineral deposits; these are of interest for estimating the ability of the geologic medium to concentrate a substance in solid, liquid, or gaseous form within bounded volumes. The screening power of clay strata is convincingly demonstrated by the formation and survival of oil and gas fields. Clay minerals have high sorption capacities for trace components of flowing liquids, such as radioactive nuclides. From this standpoint, oil and gas fields and certain types of uranium deposits merit closer attention. The laws that describe the formation and structure of petroleum and gas pools are highly useful in refining the criteria used to select appropriate geologic structures for the disposal of liquid radioactive and industrial wastes and in determining what methods must be used to explore and operate disposal sites.

2.3.1 Hydrocarbon Deposits

There is no single accepted view on how deposits (pools) of liquid and gaseous hydrocarbons came into being. Most petroleum geologists lean toward the idea that crude oil originated in organic matter buried in ancient bodies of water along with finely divided stream-borne products of the breakdown and weathering of terrestrial rocks. The notion of the organic genesis of oil has been advocated by V. I. Vernadskii and A. V. Sidorenko and further advanced by N. B. Vassoevich, A. A. Bakirov, I. O. Brod, and others.[25] A number of scientists have put forward the hypothesis that petroleum and gas hydrocarbons are inorganic in origin, formed by complex reactions of carbon and hydrogen in the Earth's mantle and core under anomalous conditions, followed by outgassing of the Earth's interior. Also worth noting are hypotheses about the origin of hydrocarbons from organic lithosphere material in processes taking place at depth and related to the development of magma chambers, high temperatures and high geodynamic pressures, and electric fields in deep tectonic faulting zones.

According to the basic points of the organic genesis concept, the primary source of hydrocarbons was organic matter resembling sapropel that accumulated with mineral sediments on the floors of warm seas, lagoons, and lakes. Upon a later general subsidence of these depositional basins, the overlying rocks forming from the sediments increased in thickness; high geostatic pressures and elevated temperatures developed. These conditions led to complex transformations of the organic matter and the enclosing rocks over long spans of time (millions of years). In these processes, liquid and gaseous hydrocarbons were generated, chemically and physically bound water was liberated, and fluids were expelled from rocks of the oil and gas source formations. Hydrocarbons in one form or the other migrated into reservoir rocks such as sands, sandstones, and limestones either by gravity (floating) or as a result of tectonohydraulic phenomena. The fluids moved through the permeable rocks until they encountered barriers formed by relatively impermeable confining rocks where conditions were favorable for the accumulation and retention of liquid and gaseous hydrocarbons.

Petroleum and gas collected in "traps." An example of a trap is a positive structure in the roof of a reservoir horizon overlain by relatively impermeable clay rocks or salts. Hydrocarbons lighter than the formation waters could accumulate in the axial part of such a structure, in a region where the permeable reservoir rocks gave way to less permeable rocks, or in a zone where the reservoir horizon was intersected by a tectonic fault acting as a barrier by virtue of the adjacent relatively impermeable rocks. Typical forms of pools in a western Siberian district are shown in Figure 3a.[26] The oil pools in the axial part of the anticlinal fold correspond to structural traps; those in the limbs, to lithologic traps. Typical of the latter type of pool is its formation where the rocks lay in monoclinal fashion, at a place where the migrating petroleum was held up when the permeability of the reservoir rocks decreased fairly slightly. Figure 3b illustrates the types of pools formed at

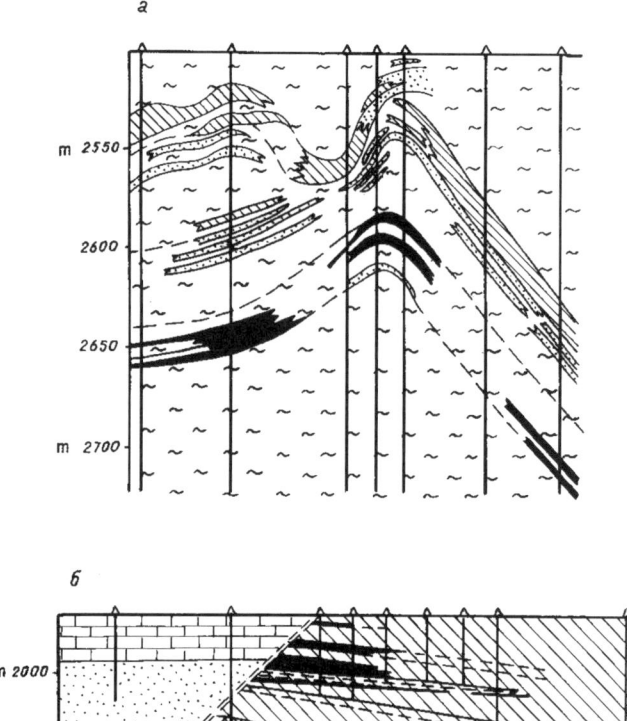

Figure 3. *Some Types of Hydrocarbon Pools in Sedimentary Strata. (a) Western Siberia, (b) Caspian Sea.*

tectonic faults where the fault plane acted as a barrier to the hydrocarbons (Dosor field, Caspian).

Tectonic traps are widely seen in the Northern Caucasus oil fields. Despite the high level of seismic activity in the region, impermeable zones of tectonic disturbances and relatively impermeable rocks adjacent to them are adequate barriers for oil and gas pools. In the Volga region at Ul'yanovsk, oil pools are also confined to deep tectonic structures. At the same time, it was possible for deep tectonic fault zones to be permeable in their early stages of geologic development, and for the contacting rocks to be jointed, so that hydrocarbons migrated through these rocks and became localized in the zone of a tectonic barrier.

Pools of hydrocarbons can form directly in the zone of the oil source rocks, in beds of porous permeable rocks that have reservoir properties and conditions to foster the accumulation of petroleum and gas. Often the pools are located tens of kilometers or more from the source rocks, such as when conditions for oil and gas collection do not exist in the immediate vicinity. The time required for pools of hydrocarbons to form, measured from when expulsion from the source rock began, is estimated to be truly geological: hundreds of thousands or millions of years.

Hypotheses of inorganic petroleum genesis call for much longer and more complex routes for the migration and transformation of the hydrocarbons. Again, however, crucial factors are the reservoir properties of permeable rocks, the isolating qualities of relatively impermeable rocks, and the conditions of occurrence.

Oil and gas pools are known to have been disrupted by a variety of processes associated with the natural transformation of the geologic medium. The remains of such pools can be identified by certain characteristic features. The disruption of a pool, especially an oil pool, is quite a slow process, estimated to require hundreds of thousands or millions of years.

Oil pools usually display high-pressure conditions, sometimes with what are considered abnormal formation pressures linked to tectonohydraulic phenomena. Even so, confining rocks of little thickness (10 to 20 m) can isolate a pool from the overlying horizons.

Among well known confining structures within the Russian Platform is the Vereian formation (Middle Carboniferous), which forms the roofs of several oil pools in Udmurtia, the Volga region at Ul'yanovsk, and elsewhere. This formation is typically 30 to 50 m thick. Oil pools are found both in terrigenous reservoirs within the formation and in carbonate reservoirs underlying it.[26, 27] According to several researchers, isolation of hydrocarbons by a confining formation requires that it have a permeability coefficient at least a factor of 400 smaller than that of the reservoir horizon.[28]

Flooding of oil-bearing formations in order to recover the oil more completely is a widely used technique in oil field development. Water is injected through specially equipped wells; the volume of injected water is often greater than the quantity of oil recovered. Flooding schemes have much in common with designs for the deep injection disposal of industrial wastes.[29]

The information given in this section cannot claim to be a detailed exposition of all problems in the formation, structure, and exploitation of oil and gas fields. It does, however, permit the formulation of some important conclusions about the deep injection disposal of liquid radioactive wastes:

- Oil pools can form within characteristic structures and local irregularities in the geologic medium associated with a decrease in rock permeability that can prevent the further migration of crude oil.

- The role of confining beds in roofing or capping pools is performed by formations of relatively impermeable rocks (clays, argillites, salts, dolomites). Thicknesses of a few meters or more are sufficient to isolate a pool for an extended period of time.

- Oil-bearing structures associated with tectonic processes, in which planes of tectonic faults act as barriers to petroleum and gas, are widespread and can isolate oil and gas pools for long periods of time. Similar structures are seen in regions of heightened seismic activity; this points to the stability of most tectonic structures after they are formed.

- The processes in which oil and gas pools are formed and re-formed involve geologically significant lengths of time.

- Reservoir rocks can enclose large volumes of crude oil and gas, as well as water injected through a system of wells in the development of oil fields.

2.3.2 Solid Mineral Deposits

Solid mineral deposits were formed when substances that had been dissolved in subsurface waters precipitated on contact with pore waters in permeable rocks. In Central Asia, there are well known infiltration (exogenetic, epigenetic) deposits of uranium that occur in bedded permeable water-bearing rocks. The ore bodies are outlined by oxidation boundaries and correspond to the region of a formerly existing geochemical barrier.[30] The deposits, which are confined to buried delta sediments of ancient rivers, are controlled by zones of stream facies. The formation of these deposits was associated with the horizontal movement of subsurface waters containing dissolved uranium compounds, which entered the waters through weathering and erosion of uranium-bearing "source" rocks located in the basins of ancient rivers. Waters enriched in atmospheric oxygen entered subsurface aquifers made of sand-clay rocks and, in the course of their movement, encountered portions of rock enriched in organic matter, which displayed a reducing character, so that poorly soluble uranium compounds were produced; these formed the secondary minerals.

The uranium ore deposits were preserved for hundreds of thousands of years in spite of the natural movement of fresh and mildly saline waters through them. The stability of uranium mineralization in permeable rocks is proven by the fact that subsurface waters of ore horizons lying at depths from tens to hundreds of meters have close to the background level of uranium. Also, the dissolution of uranium minerals requires solutions containing at least 1 to 2 g/L of sulfuric acid.

Regions of the geologic medium characterized by variations in the physical and chemical properties of the rocks or flowing solutions leading to the accumulation of an originally dissolved substance in the rocks are called physicochemical or geochemical barriers. There are several types of such barriers. An oxygen barrier forms where oxygen-free waters containing hydrogen sulfide, divalent iron, and other elements in reduced form are in contact with oxygen-bearing waters or atmospheric oxygen. Sulfur deposits are formed at an oxygen barrier. A hydrogen sulfide barrier appears where waters containing oxygen or carbon dioxide come in contact with waters containing hydrogen sulfide; such a barrier gives rise to sulfate ores of copper, lead, zinc, and other metals. A sorption barrier is created where the sorption properties change, including the ion-exchange qualities of clay rocks. Many large deposits of solid minerals developed at geochemical barriers millions or billions of years ago and have persisted down to the present.

The ability of rocks to hold up dissolved compounds also manifests itself when hydrothermal waters associated with deep magma chambers react with rocks. Hydrothermal deposits of various types form at geochemical barriers, and also where temperatures and pressures drop.

The processes by which liquid radioactive wastes disperse in, and react with, reservoir rocks clearly bear some similarities to natural hydrogeochemical processes. The accumulation of radioactive waste components in reservoir formations through physical and chemical processes can be regarded as the formation of technogenic "deposits." Formation of artificial geochemical barriers along the migration route of liquid radioactive wastes is a possible means to limit their dispersal with the subsurface waters, and also to initiate physical and chemical processes by which components

If wastes are localized at a depth of more than 400 m, the enclosing rocks will arrive at the surface after 400,000 years.

of liquid radioactive wastes are converted to solids, i.e., rocks. Research on the natural processes confirms the feasibility of such technologies.

2.4 Natural Transformations of the Geologic Medium and Tectonic Activity

The geologic structure of the Earth as a whole and of its regions is not static. The Earth is subject to changes of various types and intensities related to both deep (endogenous) processes and to the action of atmospheric factors, rivers, glaciers, and the like (exogenetic processes). In addition, human activities are becoming increasingly important as factors altering the geologic medium.

In examining the problem of deep injection disposal of liquid radioactive wastes, we must evaluate the effects of such processes on the isolation of the wastes in the subsurface. The main focus is on deep processes that express themselves in uplifting, subsidence, and horizontal displacements of the surface. These processes are linked with convective motions in the mantle, which are excited by thermodynamic phenomena as well as gravitational differentiation of mantle material and the tendency of the lithosphere to maintain gravitational equilibrium with the surface of the asthenosphere.

The downwarping of surface regions in the geosynclinal stage of development during the geologic past led to the formation of beds of sedimentary rocks up to several thousands of meters thick. The uplift of mountains was accompanied by their erosional destruction, with disintegration products being carried into downwarped areas.

Changes in the position of the surface are also observed today. According to a battery of geological and geomorphological indicators, the rate of uplifting or subsidence in regions such as the West Siberian Platform range up to 1 mm/y. Geodetic observations yield figures of 10 mm/y and more, but the movement varies in direction.[31] The discrepancy is due to the fact that geologic information reflects a general trend in the vertical movements for periods of thousands of years or more, while geodetic data relate to short-period changes in the surface position, not linked solely with deep processes, which exhibit fluctuations. The direction of vertical movements of the surface can change with a period that varies from place to place, so that uplift alternates with subsidence. Horizontal movements are manifested in plate tectonics (continental drift) and in the development of the mid-oceanic ridges, subduction zones, and seafloor spreading.

Movements of the Earth's crust are accompanied by tectonic fracturing and faulting that occur in rock strata when different portions of rock undergo variations in direction or velocity of movement. Tectonic processes are associated with sources of seismic vibrations. Tectonic faulting is seen everywhere and is marked chiefly by displacement of geologic formations and rock beds. The following questions must be considered in any discussion of the deep injection disposal of radioactive wastes and the effect of tectonic processes on their isolation.

- If uplift of the crust should occur, would wastes and the enclosing rocks reach the surface and become part of the environment of direct human activity?

- What effect would abyssal fractures or tectonic faults, both existing and newly formed, have on the isolation of the wastes in the subsurface?

- How would natural seismicity affect the wastes?

Experience in geologic observations and special studies makes it possible to answer these questions. If we assume the surface to have a constant uplifting rate of 1 mm/y, a little more than what is actually seen in the southern part of the West Siberian Platform, then the change in surface level will be 1 m after 1000 years. This much change will not lead to marked changes in geologic conditions in the period when the wastes retain their radiotoxicity. If the wastes are localized at a depth of more than 400 m, the enclosing rocks will arrive at the surface after 400,000 years, supposing that the uplifting of the surface does not alternate with subsidence in this time.

Under appropriate geologic conditions, tectonic processes will not affect the localization of wastes over realistic spans of time. In areas of recent mountain building, however, surface uplifting may be more rapid and may not be monotonic. Volcanic activity is also a safety concern in regions of vulcanism. For this reason, radioactive wastes are not buried under such conditions.

Areas in which the burial of liquid radioactive wastes is permitted in Russia usually have a two-part structure. The lower portion or basement is made up of crystalline rocks, while the upper portion is a sedimentary complex including reservoir horizons as well as confining formations. Tectonic faulting is widespread in the basement, where it was produced by deformation of the rocks at their geosynclinal stage of development hundreds of millions or billions of years ago. Some tectonic fractures also appear in the sedimentary cover; these were caused by tectonic movements while the sedimentary strata were forming or after sedimentation was complete. In the latter case, the tectonic faults extend to the surface.

Secondary or nontectonic fractures are also present. These were not formed by deep processes but resulted from movement of rocks due to gravity gradients during surface uplifting or ancient rockslides. Such faults are seen in the upper beds of sedimentary rock; they die out with increasing depth. The newest tectonic structures are those formed in the late Cenozoic (about 30 to 35 million years). Such structures are usually mirrored in the bedding of the sedimentary rocks in the upper parts of the section and also in the surface relief.

In addition to faults that have displaced the basement rocks and the later overlying sediments, "structural lines" are frequently encountered. Easily interpreted in aerial and satellite photographs as well as topographic maps, these are linear elements of the topography or landscape: valleys, edges of marshes, scarps, and distinct feet and bends of slopes. Structural lines are not, in the great majority of cases, breaks in the continuity of the sedimentary rocks with vertical or horizontal displacements. Some scientists propose that they may be zones of weakness within which an increased level of jointing is possible and that they may correspond to tectonic fractures in the basement.

Views on the role of structural lines or "lineaments" often differ. It is thought that the zones of lineaments, and especially areas where they are close together or intersect, are characterized by high permeability of the sedimentary rocks in the vertical direction. But pools of hydrocarbons in certain areas are confined precisely to intersections of lineaments and to deep tectonic fractures. In Udmurtia and the Volga region at Ul'yanovsk, for example, tectonic oil and gas traps are widespread. Obviously, an increase in permeability from the lower to the higher horizons would not promote the formation of these deposits. The class of tectonic structures called active faults are the focus of much attention. Experts differ on the criteria used to identify active faults. The term is usually understood as referring to linear, disjunct, tectonic fractures where there is evidence of movement with the following characteristics:

- consistent in direction

- having a velocity on the order of 1 cm/y

- resulting from a deep geodynamic cause

- having manifested itself in Quaternary time, or in the last 0.7 million years.

The dynamic action of the fault has naturally given rise to a destructive field, which has led to increased fracturing and permeability of the rocks. Therefore, the localization of liquid radioactive wastes in fault zones cannot be guaranteed over the long term.

Such areas are excluded from the class of suitable sites for radioactive waste burial.

Active faults are seldom seen in regions where conditions exist for deep injection disposal of liquid wastes, as in platform regions. There are several signs—geological, geomorphological, hydrodynamic, hydrogeochemical—used to identify active faults. If necessary, special studies can be performed to characterize a fault.

Some researchers lean toward ascribing active fault properties to all geosuture zones on platforms. These zones separate "geoblocks" and can be identified by the concentration of structural lines and lineaments.[32] This notion cannot be fully accepted, although some geosuture zones and portions of zones do display signs of active faults.

For the most part, tectonic faults are stable structures with no tendency toward change of any kind. According to I. V. Anan'in, reporting data on the transmission of seismic energy in strong earthquakes, only 18 to 20% of deep faults in the Caucasus have fields of increased stresses, which can serve as evidence of recent activity. The remaining faults are "transparent" to seismic energy. Fault zones in Ciscaucasia are hydrodynamic barriers, forming traps for petroleum and gas pools that have been preserved for hundreds of thousands or millions of years. Thus the presence of tectonic or abyssal faults at a proposed liquid radioactive waste disposal site does not necessarily mean that the site is unsuited to this purpose.

Most fractures in platform regions of interest for disposal are stable over tens or hundreds of millions of years, with faults confined to the basement or the lower part of the sedimentary cover, below the prospective reservoir horizons. However, tectonic fractures showing clear signs of active faults or cutting through confining formations, especially fractures that create permeable sliding planes, are bound to cause concern. It is clear that sites near such fractures (out to some kilometers away) would be inappropriate for disposal sites. The tectonic conditions for any burial site must be investigated. Figure 4 shows in schematic form the types of tectonic fractures that bear on the conditions for deep injection disposal of liquid radioactive wastes.

Natural changes in the geologic medium are accompanied by the appearance of deep sources of seismic vibrations. These vibrations manifest themselves at the surface as earthquakes, which range in severity up to destructive events. Natural seismicity is a factor limiting the creation of critical structures, including those relating to waste handling. As far as the deep injection disposal of liquid wastes (including radioactive wastes) is concerned, the seismic danger must be evaluated in a special way because seismic action decreases with depth.

Natural seismicity is a factor limiting the creation of critical structures, including those relating to waste handling. As far as deep injection disposal of liquid radioactive wastes is concerned, the seismic danger must be evaluated in a special way because seismic action decreases with depth.

	Description of fracture	How detected and investigated
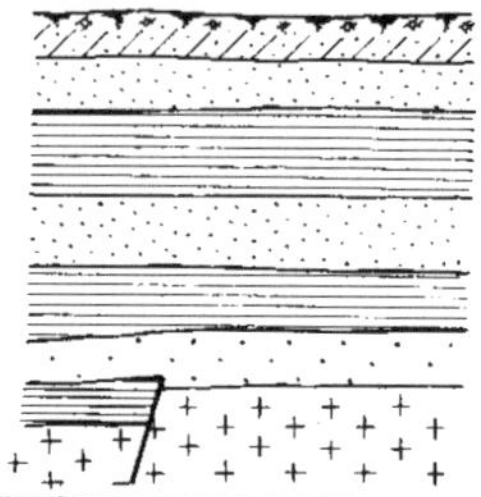	Confined to basement; weat hering crust seen. Has virtually no effect on isolation conditions of reservoir horizons.	Geophysical methods (magnetic, seismic); 2-3 test wells. Wastes can be injected in region of tectonic fracture.
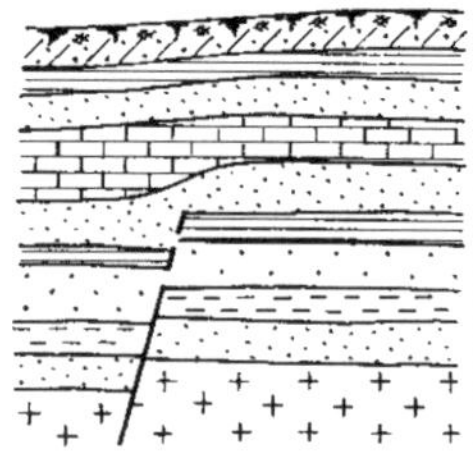	In basement and formations covering it; higher in section, seen as plicated bedding. May affect isolation in intervals broken continuity.	Geophysical methods; 2-3 test wells. Wastes can be injected if isolating qualities exist and their future survival is insured.
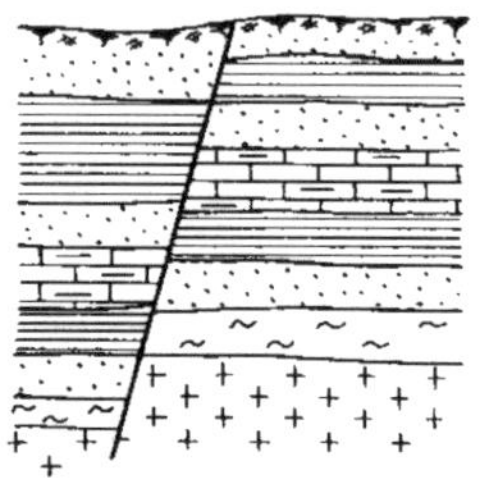	Fracture cuts basement and shallower formations; plane of movement "mudded off" and acts as barrier isolating reservoir formations.	Geophysical methods, test wells, filtration tests. Solid proof is required in order to demonsrate possibility of injection.
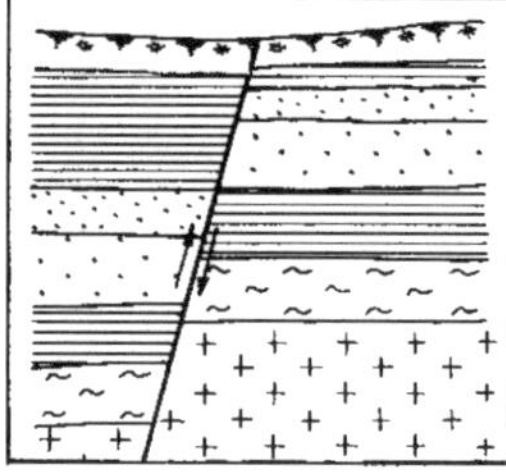	Fracture cuts basement and shallower formations; plane of movement is permeable and formations communicate along it.	Geophysical methods, test wells, filtration tests. Injection is not possible.

Figure 4. Classification of Tectonic Fractures

According to geologists studying the burial of solidified radioactive wastes for Ontario Hydro in Canada, the intensity of seismic action decreases nearly exponentially with depth. Cases are known, for example in China, where mine workings have remained stable in earthquakes, and all the miners were returned to the surface even though a populated area next to the mine was in ruins. In the destructive Gazli earthquake, deep natural gas wells suffered almost no damage, while surface equipment was destroyed.[33]

For these reasons, seismic restrictions on liquid radioactive waste disposal sites have to do mainly with surface facilities such as wellhead buildings, pipelines, and pumping stations. It may be necessary to build surface facilities using earthquake-resistant construction.

Seismic dislocations may have a marked effect on geologic formations, disrupting or sharply altering the bedding of rocks when relatively shallow-focus earthquakes take place. In seismic regions where such earthquakes occur, liquid radioactive wastes usually are not disposed of by deep injection. A local increase in vibration at the surface (relative to the predicted intensity for the region as a whole) may result from the composition, flooding, and bedding of the ground. This factor is significant for surface installations but not for deep wells or for the reservoir formations containing the wastes.

Thus, based on the literature and on observations of the nature and consequences of earthquakes, boreholes and reservoirs used for disposal of liquid radioactive wastes would show no appreciable effects. Such sites are usually located in platform and foredeep regions where seismicity is not pronounced. The surface components of storage sites would not be endangered by earthquakes either, provided the buildings are low (1 to 2 stories), foundations have short linear dimensions, and soil loads are not heavy.

Exogenetic changes in the geologic medium have to do primarily with the work of rivers, glaciers, and climatic factors. The effects of such phenomena are thought to extend to depths ranging from several meters to tens of meters, without impacting reservoir formations into which liquid radioactive wastes are injected. Human activities, however, could have a different outcome. The drilling of deep wells or mines in radioactive waste disposal sites can break into reservoirs containing wastes, allowing waste components to reach the surface or contaminate shallow groundwaters. To prevent this from happening, exploitation of the subsurface is restricted in disposal site areas. Mining licenses are granted only under strict conditions and within defined boundaries. Information on the nature and consequences of subsurface use is put in a format suitable for extended storage and for use by regulatory and supervisory agencies as they issue their decisions and licenses.

Deep reservoir formations are difficult to access, whether accidentally or deliberately. Breaking into these formations would require special drilling equipment operated by highly skilled specialists. As a result, uses of the subsurface can be effectively regulated, and there is no need to establish special protective arrangements at disposal sites after operations are concluded. Obviously, required safeguards for wastes stored on the surface are quite different.

In a hypothetical situation, information about the deep injection disposal of radioactive wastes might be lost by future generations, and a reservoir formation containing such wastes might be penetrated by accident. Clearly, such a loss of information would occur only if the entire experience of prior generations were lost, including knowledge of well drilling, mining operations, and

Drilling of wells or mines into deep disposal sites could break into reservoirs containing wastes, allowing contamination. To prevent this, exploitation of the subsurface is restricted in disposal site areas.

so forth. The recovery of completely lost technologies would require, as history teaches, no less than some thousands of years, in which time the radionuclides present in the wastes would have decayed.

Another consequence of human activity is alterations in the geodynamic equilibrium of the geologic medium through changes in the hydrostatic conditions of reservoir formations resulting from waste injection. Many specialists are aware of the seismic activity caused in this way at the Rocky Mountain Arsenal near Denver. Liquid wastes were injected directly into a permeable tectonic fault zone. The increase in hydrostatic pressure reduced the adhesion between blocks on either side of a tectonic fault plane. The blocks were in a stressed condition, and in the new situation they began to move, generating weak seismic vibrations. Low-intensity earthquakes were detected. A similar induced seismic mechanism was later confirmed by special studies in oil fields.[34, 35] In order to prevent such effects, liquid wastes should not be injected directly into the permeable zones of tectonic faults, where the contacting blocks are made up of hard rocks, or into regions of increased seismicity.

2.5 Disposal of Liquid Radioactive Wastes and Extraction of Minerals

The deep injection disposal of liquid wastes imposes restrictions on the use of the geologic medium for other purposes, including the working of mineral deposits within the area of influence of the injection. In order to cut the loss of natural subsurface resources due to waste burial, signs of minerals are followed up during preliminary studies of the region, and any promising areas are assessed. The decision to develop a burial system must take into account mineral deposits as well as shows of ore, oil, or gas. If necessary, the interaction between the buried wastes and the development of subsurface resources is calculated. For example, calculations for Udmurtia and the Volga region at Ul'yanovsk have shown that a separation of 15 km or more from oil pools is sufficient for a deep injection disposal system, even if the wastes are put into the same reservoir rock that holds the petroleum.

A valuable mineral resource is subsurface waters, whether fresh (usable for water supply) or saline (usable for curative purposes or for the extraction of iodine and bromine). Subsurface waters are encountered virtually throughout sedimentary complexes that contain reservoir beds. The deep injection disposal of wastes, including liquid radioactive wastes, inevitably excludes some such resources from possible use. Calculations have shown, however, that these losses in many regions are relatively slight, and that the beneficial effect of preventing the wastes from harming people far outweighs those losses.

In accordance with applicable standards, the use of subsurface waters located in a reservoir formation or a shallower buffer horizon within a subsurface exclusion region is forbidden (second zone of sanitary protection area). The ban on the use of waters in higher-lying horizons, including shallow ones, extends to the region where injection wells are located (first zone of sanitary protection area). At the same time, subsurface waters in these beds can be used as a source of process water for the injection system. Outside the boundaries of the exclusion region is a third zone of the sanitary protection area where the construction of water intakes must be adapted to the conditions of waste injection. Figure 5 shows a waste disposal system with its exclusion region and region of restricted subsurface water use.

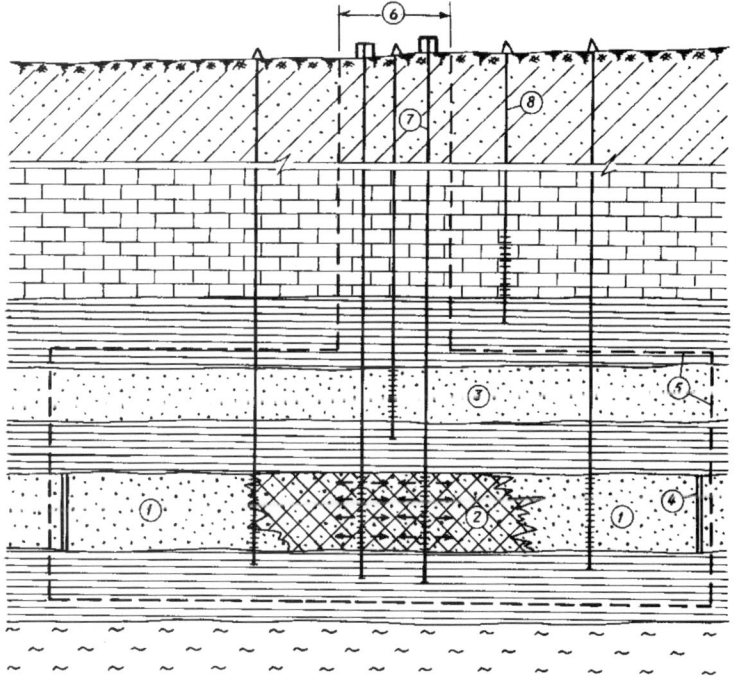

Figure 5. *Subsurface Exclusion Zone of a Liquid Radioactive Waste Disposal Site. 1 - reservoir formation, 2 - region occupied by wastes, 3 - buffer formation, 4 - predicted boundary of maximum waste disposal, 5 - boundary of exclusion zone, 6 - injection contour, 7 - injection well, 8 - observation well.*

The bulk of natural freshwater resources is concentrated in the upper part of the geologic section, mainly down to depths of 100 to 200 m; the withdrawal of the natural reserves in deep horizons will have little effect on the total reserves. Calculations have determined the loss to subsurface reserves in the 1500-km² water management region in the center of which liquid radioactive wastes from the Siberian Chemical Combine will be buried. The reserves of subsurface waters that cannot be used as a consequence of the disposal site will amount to 1 to 3% (depending on the computational method). The boundaries of the region were based on economic links between existing and possible users of subsurface water as well as an administrative map.

3 Investigation and Justification of Deep Injection Disposal of Liquid Radioactive Wastes

3.1 Concepts and Their Realization

Solving the problem of the deep injection disposal of liquid radioactive wastes proved to be a complicated scientific and engineering task involving multiple research, analysis, and design efforts. The first steps toward dealing with the problem were to establish conceptual foundations and to institute comprehensive studies, geological exploration, and preliminary design.

The first concrete proposals to begin work on the underground disposal of radioactive wastes were submitted to the government in 1957 by Prof. S. A. Voznesenskii of the Ural Polytechnic Institute. The occasion for this proposal was, obviously, the explosion of a storage vessel for high-level wastes in the southern Urals in that same year. The facility contained 20 million Ci of wastes, and the accident contaminated the Techa River, adjacent territory around the surface storage areas, and other bodies of water. Having taken part in assessing the environmental effects in the vicinity of the southern Ural radiochemical plants [now Production Association (PA) Mayak], Voznesenskii had seen first hand the danger posed by the technologies then used for handling liquid radioactive wastes.

Voznesenskii's proposals found supporters in Academicians V. I. Spitsyn, a renowned physical chemist, and A. P. Vinogradov, who headed the Academy of Sciences Institute of Geochemistry and Analytical Chemistry. From preliminary studies, they were aware of the great sorption capacities of rocks for nuclides in wastes; they also knew of the U.S. experience with disposing of liquid radioactive wastes in surficial deposits at Hanford, where again the dry alluvium was able to delay nuclide dispersal. In contrast with the approaches used in America, however, petroleum geologists N. A. Kalinin and M. N. Baranov suggested using deep reservoir formations for waste disposal because these formations were positively isolated from the surface by layers of clay rocks.

Deep injection of waters produced from oil fields had already been carried out, and nonradioactive liquid wastes were already being injected in deep reservoir formations in the U.S., Germany, and elsewhere. Liquid radioactive wastes were not being injected outside the USSR at that time, because of the lack of appropriate geologic conditions at U.S. atomic plants and research centers (e.g., Hanford, Savannah River, Oak Ridge).

The first concrete proposals to begin work on the underground disposal of radioactive wastes were submitted to the government in 1957 following the explosion of a storage vessel for high-level wastes in the southern Urals in that same year.

Scientific and technical proposals relating to the conditions and procedures for deep injection disposal of liquid radioactive wastes were examined and approved by the Scientific and Technical Council of the Ministry of Medium Machinery Construction, with Academician I. V. Kurchatov as chairman.

The principles of deep injection disposal accepted in 1957 are summarized below:

- It is possible, in theory, to use several types of geologic formations for the deep injection disposal of liquid radioactive wastes.

- The deep injection disposal of liquid radioactive wastes can be carried out only under certain geologic conditions, and a number of requirements apply.

- Deep injection disposal must be preceded by study of the geologic medium and the wastes; enough scientific data should be acquired to ensure the consequences of the injection project can be dependably forecast.

- The wastes to be injected must be compatible with the geologic medium.

- The disposal procedure and any processes taking place in the subsurface must be monitored; injection of wastes must alter the natural conditions of geologic formations to the minimum possible degree.

These principles are the foundation for requirements regarding the geologic medium and the technology of deep injection disposal: localization of liquid radioactive wastes within definite boundaries, compatibility of wastes, and careful monitoring of all processes. Localization corresponds to a basic tenet of mining law, that the affected region must be limited to the duly created exclusion zone.

Clearly, not every geologic formation was suitable for the disposal of liquid radioactive wastes. Formations must have enough capacity—after injected wastes have filled the pore space of the rocks, they should occupy only a relatively small area of the reservoir formation. Formations suitable for disposal must be isolated from the surface and shallow subterranean waters by relatively impermeable strata with confining properties. Above a reservoir bed and separated from it by a confining bed, a buffer horizon must exist that also meets the requirements for reservoir formations, so that it can afford standby capacity between the reservoir bed and shallow horizons.

The region possibly affected by the disposal project must not contain zones where the relatively impermeable confining rocks pinch out, regions of lithologic substitution, or permeable fault zones where leakage could occur. The velocity of the natural movement of subsurface waters in the reservoir formation must be small enough that the decay of nuclides to safe levels takes place a fairly short distance from the disposal site; estimated velocities must be a few meters per year.

The deep injection disposal of liquid radioactive wastes must not be undertaken close to a discharge area of the reservoir formation into shallower horizons. Deep injection disposal must not interfere with development of mineral deposits if any are present in the region.

Wastes destined for injection must be compatible with the geologic medium. Their injection must not irreversibly change the reservoir formation. Examples of such changes include lowering the flow capacity so as to block deep injection disposal, intense heating and gas evolution in the formation, and dissolution of rock-forming minerals.

Waste injection practices must not upset the geodynamic equilibrium of the geologic medium; this standard is met by limiting the injection pressure. Because the main engineered features, drilled wells, ensure the reliable isolation of the reservoir formation from all higher-lying horizons, they must be monitored and in a state of good repair while in use; they must be sealed after they are taken out of operation. Deep injection disposal is a monitorable process; that is, needed information about the behavior of the wastes in the subsurface, their location, and any processes going forward must be available at all times.

To satisfy these geologic and technical requirements, data must be obtained on the geologic structure and properties of the subsurface medium at proposed burial sites, on the characteristics of the wastes, and on the interaction of wastes with geologic medium. The specialists creating disposal systems must have access to needed hardware and methods of investigating all questions relating to deep injection disposal. The volume and quality of these data must be such that practical decisions can be based on adequate grounds, and the consequences of disposal operations must be predictable with fair confidence. The basic conceptual points on deep injection disposal of liquid radioactive wastes are presented in Table 4.

A point of concern in the conceptual analysis, with reference to new environmental and public-health standards, has been the responsibility to later generations, who will to one degree or another face the consequences of these activities. It has been maintained that subsurface regions containing wastes will be similar to deposits of minerals, including radioactive minerals, which human beings encountered long ago and have managed to handle safely.

Deep injection disposal is a monitorable process; that is, needed information about the behavior of the wastes in the subsurface, their location, and any processes going forward must be available at all times.

The consequences of injecting liquid radioactive wastes will be orders of magnitude less detrimental than the consequences of releasing the wastes into pools, streams, and reservoirs in the Earth's surfaces.

Table 4. Principles Governing Deep Injection Disposal of Liquid Radioactive Wastes

Properties of Geologic Formations	Waste Compatibility with Geologic Medium	Disposal Technology
Limited volumes of liquid radioactive wastes can be placed in deep porous rocks (reservoir formations).	Placement of liquid radioactive wastes in geologic formation does not promote processes hindering burial or reducing degree of isolation of wastes.	Injection practices do not upset geodynamic stability of geologic medium, violate continuity of reservoir formation, or lead to induced seismicity.
Wastes are isolated within set boundaries by virtue of low permeability of rock beds overlying reservoir formations (confining beds) and low velocity of natural water movement in reservoir formation.	Interaction of liquid radioactive wastes with rocks and subsurface waters causes radioactive nuclides in wastes to go into solid phases (rock-forming).	Design and technology of wells ensures that all horizons penetrated by wells remain isolated from reservoir forma-tion. Wells are sealed when operation is terminated.
Geologic formations suitable for deep injection disposal of liquid radioactive wastes must meet certain requirements.	Deep injection disposal of liquid radioactive wastes leads to formation of deposits of waste components; these are technogenic deposits.	Boundaries of subsur-face exclusion zone and sanitary protection zones are defined in the region of the deep disposal site. Use of the subsurface is restricted within these zones.
Waste disposal opera-tions are preceded by study of geologic formations.		Processes of waste injection and migration in geologic medium are monitored.
Geological and ecological consequences of waste disposal are predicted with a sufficient degree of confidence.		

The consequences of injecting liquid radioactive wastes will be orders of magnitude less detrimental than the consequences of releasing the wastes into pools, streams, and reservoirs on the Earth's surface (though there was no alternative from the 1960s to 1980s). Wastes stored at the surface inevitably cause radiation exposure to the population, leading to genetic defects whose correction or neutralization will take many years if at all possible. The efforts and expenditures required of our descendants to deal with such wastes will be far greater than the costs due to limited use of the subsurface at disposal sites.

Another point of concern has been the issue of localization: for how long can localization of deep injected wastes be guaranteed? It was first thought that localization was governed by the time required for fission-product nuclides to decay to safe levels: from 300 to 1000 years depending on the concentrations and composi-tion of the nuclides.

However, this time factor for waste localization was later re-evaluated because of the presence of stable components in the wastes (such as salts) that are also pollutants, together with unrecoverable trace amounts of long-lived nuclides. If the localization time is linked with the progress (or, on the other hand, the regress) of society, then the needed time can be estimated at between 1000 and 10,000 years.

The question of safe levels of radionuclides (as waste components) in the geologic medium has not escaped attention. The first assumptions about safe, allowable levels of nuclides in drinking water were based on the existing health-physics standards. Because these standards may vary over time, it was later suggested that the safe level should be set at the level of the variability of the natural contents of uranium, radium, thorium, and ^{40}K in rocks and waters (aside from ore concentrations), adjusted for the relative biohazard presented by these fission-product nuclides and by natural radioactive elements.

After the government approved the principles discussed above, the All-Union Scientific Research and Design Institute of Industrial Technology (VNIPIpromtechnologii) created Specialized Scientific Research Laboratory No. 5 (NIL-5) in June 1957, pursuant to a decree of the Ministry of Medium Machinery Construction (now Minatom) (Directive No. 317s, June 4, 1957). The task of this unit was to conduct research, develop design and budgeting documents for the construction of deep injection disposal sites for liquid radioactive wastes, and introduce the new technique for radioactive waste handling at enterprises in the sector. The laboratory was headed successively by N. A. Kalinin, M. K. Pimenov, F. P. Yudin, and A. I. Rybal'chenko. In order to study and investigate special topics, the laboratory brought in specialized scientific research, production, design, and planning organizations affiliated with the Academy of Sciences, Gosstroi, Mingeo, Minzdrav, Minvuz[a] and other agencies.

In 1958, at the suggestion of the Ministry of Medium Machinery Construction and the Ministry of Geology, the Soviet government issued Directive No. 3019rs, dated September 13, 1958 and signed by A. N. Kosygin, dealing with geological exploration and other studies on the deep injection disposal of liquid radioactive wastes at the Siberian Chemical Combine (Tomsk-7), the Mining and Chemical Combine (Krasnoyarsk-26), the Mayak Combine (Chelyabinsk-65), and the Scientific Research Institute of Nuclear Reactors (Dimitrovgrad).

(a) In order: State Construction Council of the Council of Ministers; Ministry of Geology; Ministry of Public Health; Ministry of Higher and Secondary Special Education.

The effort to create systems for deep injection disposal of liquid radioactive wastes involved specialized organizations within the Ministry of Geology (now Minprirodresurse), which carried out geological explorations under the leadership of K. J. Antonenko and, later, Yu. S. Tatarchuk, B. V. Grafskii, and E. I. Chapovskii. The Institute of Physical Chemistry of the Academy of Sciences performed many studies of wastes, rocks, and their compatibility. This work was accomplished under the direct leadership of the institute's director, Academician Spitsyn, and V. D. Balukova, head of the specialized laboratory.

Occupational and radiation safety became a major concern, and to this end environmental impact assessments were done for the burial of liquid radioactive wastes. These assessments included forecasts of the consequences, and determinations of the health aspects, of deep injection disposal. These studies were carried out by organizations of the Ministry of Public Health and Institute of Biophysics under the direction of A. S. Belitskii and A. I. Ryzhov.

Led by Chief Engineer E. D. Mal'tsev, M. K. Pimenov, and F. P. Yudin, the All-Union Scientific Research and Design Institute of Industrial Technology (VNIPIpromtechnologii) did the justifications, design, and prediction, and consequence assessment for specific installations at the Siberian Chemical Combine, the Mining and Chemical Combine, and the Scientific Research Institute of Nuclear Reactors (NIIAR).

The creation and operation of disposal systems were directly supervised by the government or by delegated state agencies. Supervisory authorities enlisted leading specialists from a variety of agencies to assess the work, set up evaluating commissions, and hold inspections at the Scientific and Technical Council level. The design documents used the term "range" for disposal systems and facilities and "deep injection storage" for the reservoir formation. In what follows, these terms will be used along with those already introduced.

3.2 Geological Exploration

3.2.1 Siberian Chemical Combine (Tomsk-7)

Geological exploration began at the Tomsk-7 site of the Siberian Chemical Combine in the southwest part of the West Siberian Basin. The central and northern portions of this basin were well known as promising oil and gas regions. The geologic section contains reservoir beds, chiefly represented by sand rocks, as well as confining beds, and strata of clay rocks. Data on file indicated that the sand-clay sediments in the plant region were several hundred meters thick, lying on a crystalline basement, and were represented by Cretaceous and Paleogene sediments. Information

about the composition and the reservoir and flow properties of
these beds, however, was limited.

The Novosibirsk Territorial Geological Administration, which
later became part of the All-Union Hydrogeological Trust, (subsequently the Second Hydrogeological Administration, now GGP
Hydrospetzgeologiya) began exploration in 1958. This work
was done by graduate geology students A. T. and R. A. Larchenko,
N. N. Tishchenko, A. I. Gorbunov, and E. R. Makarov of the
Tomsk Polytechnic Institute, under the direction of leading
specialists B. N. Savvin, N. A. Titov, and V. M. Danilovich of GGP
Hydrospetzgeologiya. The newly created geological service of the
Siberian Chemical Combine detailed M. N. Baranov, V. P. Solopov,
and L. F. Novoselov to participate in the studies.

In the methodology adopted to investigate the oil regions, the
first stage of the geological exploration was electrical and seismic
prospecting to measure the depth of the basement and its relief
and arrive at a provisional subdivision of the section. The results
confirmed that thick sedimentary strata were present, presumably
including reservoirs and confining beds of regional extent; the
beds were found to dip gently northwest. The basement was
only slightly displaced by faults.

Discussion of these results led to the decision that more detailed
and thus more costly studies should be performed, including
hydrogeological mapping at a 1:50,000 scale, drilling of wells
for a series of geophysical measurements, acquisition and examination of rock specimens, and filtration tests. Table 5 will convey
some idea of the scope and nature of this geological exploration.
The methodology of the research conducted to justify the deep
injection disposal of liquid wastes has been described elsewhere in
considerable detail [36, 37]; a summary of the work will suffice here.

The exploration made it possible to describe in detail the stratigraphic section, identify reservoir and confining horizons, plot
a hydrogeological scheme for the region, and determine the
pressure regime of aquifers and the flow gradients and natural
velocities of subsurface water flow (velocities of 3 to 5 m/y were
found). The analysis culminated in the preparation of correlation
diagrams and geological-geophysical sections, as well as horizon-
by-horizon structural maps of various characteristics (total and
effective thicknesses of beds, permeability, depth to roof and floor).
Mineralogy, porosity, flow properties, and static treatment results
were tabulated. The major results of the geological exploration
are described more fully in Chapter 5, where geological diagrams
and sections are presented for the respective liquid radioactive
waste disposal sites.

Special attention was devoted to the hydraulic conductivity properties of the reservoir horizons and their isolation from shallower
aquifers. Extended tests were performed in which researchers

*Special attention
was devoted
to the
hydraulic
conductivity
properties of
the reservoir
horizons and
their isolation
from shallower
aquifers.*

Table 5. Geological Exploration to Justify Deep Injection Disposal of Liquid Radioactive Wastes

Entry No.	Enterprise, Site	Performance Period	Types and Amounts of Work Performed							Cost During Perf. Period, 1000s of Rubles
			Seismic Prospecting, km²	Electrical Prospecting, km²	Hydrogeological Survey, km²	Test Wells, Lineal m	Number of Test Wells	Pumping Tests (single and cluster)	Injection Tests (single and cluster)	
	I. Siberian Chemical Combine									
1	Pilot Site, Area 18									
2	Experiment, site 18a	1957-1965	800	800	560	80,000	128	47	18	10,402
3	West section of Area 18a									
	II. Krasnoyarsk Mining/Chem. Combine									
4	Severnii site (horizon I)	1959-1963	1,520	1,520	150	35,000	77	12	10	3,210
5	Severnii site (horizon II)									
6	Area XXVII-1	1964-1969	—	—	180	45,000	66	8	4	3,369
7	Area XXVII-2									
	III. Mayak Combine									
8	Techa-Brodskaya syncline	1961-1965	350	350	350	37,000	47	5	1	4,557
9	Arkhipovskii graven (Uksyanskoe area)									
	IV. Sci. Res. Inst. of Nuclear Reactors									
10	Pilot site	1962-1970	200	—	—	22,000	15	5	5	3,949

injected water and monitored piezometric levels of the subsurface waters in the horizons under test, as well as in higher-lying horizons separated by relatively impermeable clay rocks. Injection was continued for 1 to 2 months at a rate comparable to a waste injection rate, so that fairly representative data were obtained. By way of example, Figure 6 shows the setup for injection tests used to determine the parameters of reservoir formations and the confining formations overlying them.

One important inference from the hydrodynamic data was that the water-saturated Cretaceous complex in which the promising reservoir horizons II and III were located did not in fact communicate with the Upper Cretaceous and Paleogene sedimentary complex in the potential area of influence of waste injection (the

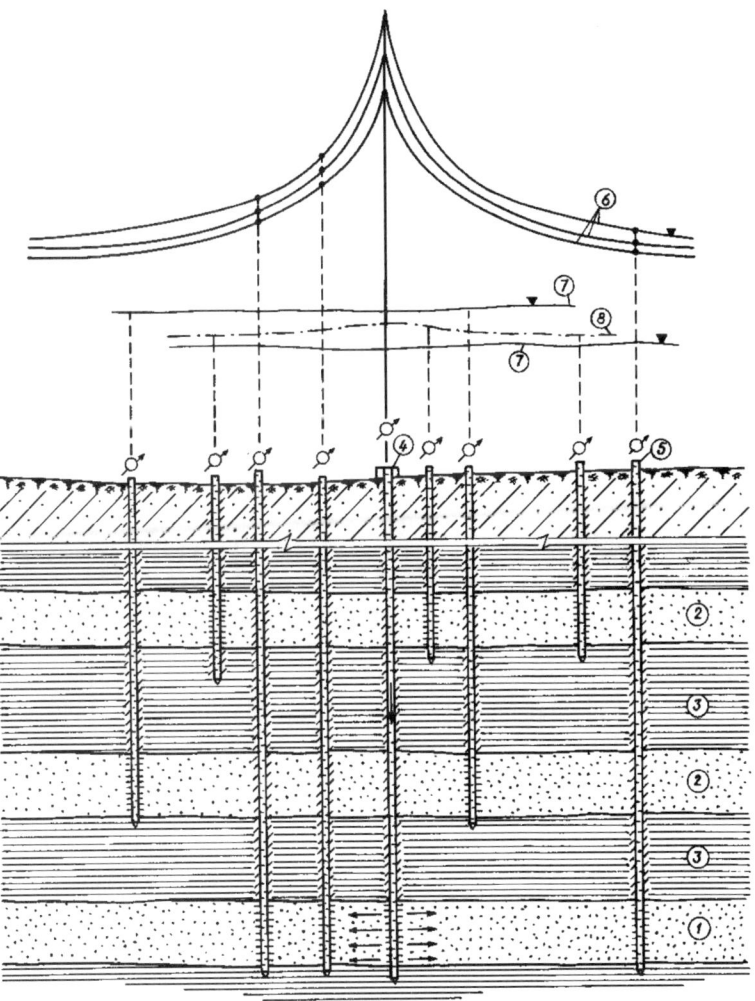

Figure 6. *Diagram of Filtration (Injection) Tests. 1 - reservoir formation, 2 - buffer and roof horizons, 3 - relatively impermeable rocks, 4 - central injection well, 5 - observation wells, 6 - pressures in reservoir formation plotted at various times (repressuring dome), 7 - level of subsurface waters in shallower horizons, 8 - variation in level if formations were in communication.*

relatively impermeable "D" horizon lay between these complexes). This finding was indicated by prolonged injection tests, measurement of helium concentration fields, and differences in the subsurface water composition.

A portion of the "bench" adjoining the disposal site on the east and presumably corresponding to an abyssal fault was studied. Data from test wells, which were spaced closer together here, showed plicated flexure-like bedding (with no break in continuity) of the reservoir and confining beds forming the structure. Experimental injection runs showed that the confining formations in this area retained their isolating qualities.

Participating in the hydrodynamic tests were specialists in oil field investigation and development at the All-Union Scientific Oil and Gas Research Institute, then the leader in the field. The vertical nonuniformity in flow properties of the beds was studied by injecting solutions of radioactive tracers into wells. Radioactive logging yielded a permeability profile for the horizons under study. Attention was appropriately devoted to examining the geological and engineering conditions in order to select the optimal well design and the optimal technologies for drilling, casing, and completion. Instrumentation and techniques were devised for monitoring the condition of the wells.

The results of exploring the Tomsk-7 site were reported. Theoretically, deep injection disposal of liquid radioactive wastes was considered feasible for that region. The results, which were scrutinized by organizations of the Ministry of Geology and by outside experts, were assessed as favorable for practical use, that is, for justifying and designing ranges where liquid radioactive wastes would be stored at the Siberian Chemical Combine.

3.2.2 Mining and Chemical Combine (Krasnoyarsk-26)

Geological exploration near the Mining and Chemical Combine at Krasnoyarsk-26 was begun later in 1958 than at the Tomsk-7 site to gain the benefit of experience for planning and for determining which studies could be omitted in the preliminary stages. The special geologic conditions in this region included sand-clay sediments containing reservoir and confining formations mainly dating from the Middle and Upper Jurassic and deposited in a synclinal structure (trough) formed by basement subsidence. The synclinal structure was bounded on the west by a tectonic fault with a 200 to 250 m offset in the basement. It was later found that the plane of the fault acted as a hydraulic barrier between the upthrown and downthrown sides.

These conditions—the synclinal bedding and the boundedness of the structure in plan—were very favorable for disposal of wastes having greater density than the subsurface waters in the reservoir formations. On the other hand, because the region was confined to the Siberian Platform/West Siberian Basin zone of articulation, and a tectonic fault was present in the area of influence of the disposal operations, detailed studies of the geology had to be performed.

Geological exploration was begun in 1958 by a party from the Krasnoyarsk Territorial Geological Administration (geologists A. V. Goncharov, A. V. Nosukhin, V. T. Ryzhenkov, and others). This party formed the nucleus of a specialized geological enterprise headed by M. M. Polyakov and subordinate to the GGP Hydrospetzgeologiya. Geophysical studies were directed by G. P. Ponsui-Shapko and I. T. Gavrilov.

In general, the methods were the same as used at the Siberian Chemical Combine. The isolation of horizons I and II (considered possible reservoirs) by higher-lying clay confining beds was confirmed by extended pumping and injection tests in which the piezometric surface was monitored in all penetrated formations; the helium content was also determined. Also, special studies were done to establish the barrier properties of a tectonic fault plane. Specialists from Moscow State University, the All-Union Scientific Research Institute of Hydrogeology and Engineering Geology, and other institutes and organizations were called upon to rectify problems that arose during the geological exploration. The exploration work was not restricted to the synclinal structure but extended to the left bank of the Yenisei River, where the sedimentary formations of the West Siberian Basin's border increased sharply in thickness.

From a safety standpoint, one part of the synclinal structure was judged most suitable for deep injection disposal. The closed character of the structure led to a recommendation that pressure in the reservoir horizons be relieved by pumping out formation water simultaneously with waste injection. The locations of the injection and relief wells had to be chosen to preclude contamination of the pumped waters. Tests performed in the course of exploration confirmed that this method of disposal was possible. The results and conclusions of the geological exploration were approved by leading specialists in the Ministry of Geology and judged suitable for design work.

3.2.3 Scientific Research Institute of Nuclear Reactors at Dimitrovgrad

The geology at the third site, the NIIAR at Dimitrovgrad in the Ul'yanovsk oblast, was radically different from that at the first two sites. Because the explored area was confined to the oil region of the middle Volga, the reservoir properties of the deep horizons (over 1000 m) containing saline waters (brines) had already been established. The section contained an identified and fairly well studied regional confining formation of Vereian sediments, which were seen practically everywhere in the Russian Platform.

The purposes of the geological exploration were to identify the most promising reservoir horizons below the Vereian sediments, confirm their isolating qualities, determine the capacities and flow properties of the reservoirs, and study the drilling conditions from an engineering geology standpoint. To examine the structural features of the region, seismic prospecting was done. This made it possible to derive structural diagrams of the reflecting and refracting horizons, which characterized the bedding of both reservoirs and confining beds.

Test wells and a combination of geological and geophysical studies made it possible to identify the stratigraphy of the section and achieve a lithologic subdivision. Two reservoir horizons were identified: terrigenous sediments of permeable zone III (depth 1440 to 1550 m) and carbonate sediments of permeable zone IV (depth 1130 to 1410 m) of the Middle Carboniferous. The presence of the Vereian confining horizon in the section was confirmed, as were its confining properties. Pumping and injection tests showed that the identified horizons could be used for the disposal of liquid radioactive wastes. The results of the geological exploration were also approved by the Ministry of Geology.

Subsequent exploration of three areas at Dimitrovgrad was supplemented by studies of the helium concentration fields and the isotopic makeup of subsurface waters. Geophysical investigations confirmed the prior conclusions regarding the geology.

3.2.4 Mayak Production Association (Chelyabinsk-40)

Geological exploration aimed at justifying the deep injection disposal of liquid radioactive wastes was also performed in the regions of the Mayak Production Association (PA) at Chelyabinsk-40 and the nuclear power plants at Ignalina, in the Kalinin and Smolensk oblasts and at Novovoronezhskii. The geology at Mayak proved unfavorable for the deep injection disposal of liquid radioactive wastes. Karsted limestones were found in the sedimentary complex of the Techa-Brodskaya Syncline, and their isolation from shallower horizons of subsurface waters was not demonstrated. The nearest suitable area was 150 km east of Mayak, in the transition zone to the West Siberian Basin (the settlement of Uksyanskoe in the Kurgan oblast). Here the reservoir horizons are confined to Triassic sand-clay strata filling a synclinal structure in the basement (thought to be formed by an ancient river bed). Geological exploration of this area formed the basis of a technical and economic justification for the deep injection disposal of liquid radioactive wastes. But the site was not selected for development of a disposal system because the area was too far from the Mayak plant.

Geological exploration in the areas of the Ignalina and Smolensk nuclear plants revealed geology suitable for deep injection of liquid radioactive wastes, and technical and economic justifications were prepared. Because of the limited waste volumes at these plants, and the antinuclear movement of the late 1970s and early 1980s, however, it was decided to solidify the radioactive wastes from the plants. Positive results of exploration were also obtained for the Shevchenko energy plant (Mangyshlak oblast), but deep injection disposal was not developed there for a variety of reasons.

Operation of the Kalinin nuclear plant originally included plans for deep injection disposal of liquid radioactive wastes, but a decision was later made to solidfy wastes instead. The wastes that were considered for burial were 1000 to 1500 m^3/d of saline solutions from chemical water treatment, which were formerly discharged into surface streams. A deep injection disposal facility was set up in 1990-1993 on the basis of the design documents prepared.

The results of exploration near the Novovoronezhskii nuclear plant were inconclusive regarding isolation of the sandy reservoir horizons, which lie on the flank of the Voronezh crystalline massif. A possible reservoir formation in the region of the Leningrad atomic power plant was the Gdovskii horizon, a reserve source for the water supply system of the oblast center, which discharges beneath the floor of the Baltic Sea. In light of these facts, it was decided not to undertake the burial of liquid radioactive wastes from the Novovoronezhskii and Leningrad plants.

When the new RT-2 radiochemical plant was built at the Mining and Chemical Combine during 1980-1985, an area on the left bank of the Yenisei River was re-explored. Exploration methods included electrical and seismic prospecting, a battery of tests performed in drilled wells, isotopic geochemical studies, and extended pumping and injection tests. Positive results were obtained, and sandy reservoir horizons isolated from the surface were identified in depth intervals of 660 to 1000 m. At the recommended site, Area 27, the main recommended bed (Minzhul'skii horizon, Middle Jurassic) is confined to a belt 3 to 4 km wide in the formation, bounded on the southwest and northeast by zones of lithologic replacement of sands by clays. There are thick strata of relatively impermeable clay rocks with surface up to the top of the first reservoir formation, which lies at depths of about 660 m.

These geologic conditions, the low velocities of the natural movement of subsurface waters, and the high sorption capacity

The geologic formation selected for waste injection must be isolated from the surface and from shallow-lying fresh waters that may be candidates for water-supply use.

of the reservoir rocks, imply that injected liquid radioactive wastes would be fairly reliably localized. Nevertheless, the wastes would have to be delivered to the disposal site by pipeline, which in turn would have to be placed in a tunnel beneath the Yenisei River. These facts elicited public protests by people who doubted the safety of such a transportation system, thus a decision was made against creating a disposal range.

3.2.5 General Requirements for Geologic Formations Based on Exploration

The experience gained in conducting geological exploration and discussing the results, as well as the subsequent steps of justification, design, construction, and operation of disposal systems, have made it possible to state and refine general requirements on geologic formations suitable for the disposal of liquid radioactive wastes. The first and most important requirement is that the geologic formation selected for waste injection must be isolated from the surface and from shallow-lying fresh waters that may be candidates for water-supply use. This isolation requirement can be extended to mineralized waters of possible curative value if such waters are present. Isolation must be ensured not only vertically but also horizontally (isolation from contiguous portions of the selected formation).

This requirement will be met if the geologic formation (reservoir formation) is overlain by a sufficiently thick layer of relatively impermeable rocks, in particular rocks of clay composition, and if the velocity of the natural movement of subsurface waters in the reservoir formation is low. Quantitative measures are set and refined on a case-by-case basis. Experience to date indicates the following: the thickness of the relatively impermeable formations overlying the reservoir formation must be no less than some tens of meters, the rocks must have a hydraulic conductivity of 10^{-3} to 10^{-4} m/d or less, and the subsurface waters must naturally move no faster than 10 m/y.

A crucial factor for waste isolation is the interaction of the wastes with the geologic medium, since interaction processes lead to the conversion of liquid waste components to the solid phase (as rocks). Other important considerations are the presence of a buffer horizon above the reservoir formation, the continuity of the confining bed or formation, the absence of lithologic windows and permeable fault zones in the confining bed, and the flow uniformity of the reservoir formation (i.e., freedom from zones of anomalously high permeability).

The next most important requirement is that the reservoir formation have high enough capacities that sufficient volumes of wastes can be placed within limited regions. The specific capacity is evaluated by the volume of wastes that can be placed in it per unit area of the reservoir formation. The specific capacity is the product of the effective formation thickness and the effective porosity. The effective thickness is defined as the sum of the permeable beds in the reservoir formation filled with wastes; the effective porosity is the fraction of the rock volume comprising open pores that can be filled with wastes. The specific capacity for the liquid phase (components in solution that do not interact with the rocks) is estimated at 0.5 to 10 m^3 of wastes/m^2 of formation (or 5000 to 100,000 m^3 of wastes per hectare of formation), corresponding to a thickness of 10 to 100 m and a porosity of 0.05 to 0.2.

Another important requirement is the possibility of confidently determining the geologic structure and the properties of the geologic formations as a basis for realistically forecasting how the wastes will behave in the subsurface and what consequences will follow from disposal operations. This requirement is satisfied for geologic formations where the properties and characteristics vary in a monotonic way: sand-clay sedimentary formations with uniformly distributed porosity in which tectonic structures are weakly developed and basically plicated in nature. Some types of carbonate reservoirs with primary and secondary porosity also meet this requirement. The geologic sections of such formations as directly determined at discrete points (test wells) correlate among themselves to a satisfactory degree, so that the characteristics obtained can be interpolated. Disturbances of the piezometric surface of the subsurface waters created by pumping and injection tests can disperse over large areas; this means that representative data can be obtained. At the same time, hard rocks with a transforming fracture type of porosity and vuggy carbonate rocks are very difficult to study, and the data obtained in these media contain large uncertainties.

A further necessary condition is that the construction of adequately reliable wells must be possible. The development of karst in the upper part of the section may cause frequent failures in well construction and may also prevent cement from rising as high as intended during construction. In this case, the reservoir formation would not be adequately isolated from shallower horizons. These basic requirements imply several others, among them location of reservoir formations in the hydrodynamic zone of impaired water flow, and hydrochemical zonation of the stacked beds. Table 6 lists the principal requirements on geologic formations for the disposal of liquid radioactive wastes, along with characteristics of the geologic medium and methods of studying them.

Table 6. Requirements of Geologic Formations Suitable for Disposal of Liquid Radioactive Wastes

General Characteristics of Formations	Governing Parameters; Ranges of Values	Methods of Investigation
Isolation of reservoir horizons from shallow freshwater horizons by relatively impermeable rocks with confining qualities.	Filtration factors of confining beds, 10^{-3} to 10^{-4} m/d; thickness 30-40 m or more; absence of lithologic and tectonic flow "windows" in area of influence of disposal.	Test wells for geophysical studies; examination of core specimens; filtration tests; hydrochemical studies (helium survey, isotope determinations, hydrochemical characterization of subsurface waters).
Low velocities of natural movement of subsurface waters in reservoir formation.	Natural velocities not over 10 m/y.	Determination of piezometric surface of subsurface waters by downhole observations; determination of flow properties from filtration tests.
Adequate capacity of reservoir formation.	Specific capacity for liquid phase > 0.5 m^3 wastes/m^2 of formation; effective thickness 10-200 m; porosity > 0.05; rocks with sorption (holdup) capacity for nuclides.	Combination of downhole studies and core specimen examination; filtration and migration tests.
Presence of buffer horizon between reservoir formation and shallow freshwater horizons.	Thickness of horizon > 10 m; porosity > 0.05 m; flow properties similar to reservoir formation.	Similar to "Isolation" entry.
Absence of relief zones of reservoir horizons and of active faults in potential area of influence of deep injection disposal.	Such zones and structures must be absent within a radius of 10-15 km from disposal site.	Area geophysical studies; hydrogeochemical studies; examination of neotectonic structure; analysis of geological data in many aspects.
Absence of mineral deposits in area of influence of deep injection disposal.	Oil and gas fields confined to horizons to be used for burial must be no closer than 15-20 km from range; those lying in other horizons, no closer than 10-15 km.	Thorough analysis of geological data.

3.3 Physicochemical Studies of Wastes and Their Interaction With the Geologic Medium

In the field of deep injection disposal, the focus of physical and chemical research is determined by forecasting the variations in physicochemical and geochemical conditions when pore spaces in the formation are filled with wastes. When liquid radioactive wastes are introduced into a reservoir, the natural geochemical equilibrium is upset; physical and chemical reactions ensue that lead to changes in the compositions of the liquid and solid phases. The acidity or alkalinity of the pore solution is altered most easily. In the final analysis, any starting solution tends to come to equilibrium with the formation conditions.

An important factor determining the possibility of disposal is the nature of the physical and chemical processes involved in reaching new geochemical equilibria. The physicochemical properties of the wastes, their chemical and radiochemical composition, and the change in their properties when filling the reservoir are very significant. The physical and chemical properties of the wastes must guarantee stable operation of injection wells, prevent colmatage of reservoir zones adjacent to the filter zone, promote the transfer of contaminants and radionuclides to the solid phase, and retain migrating waste components by the pore-filling medium of the reservoir.

Recommended studies to resolve physical and chemical questions connected with deep injection disposal of liquid radioactive wastes include determination of the following:

- chemical composition and physicochemical properties of wastes and rocks

- behavior of waste components during interaction with formation waters and rocks, including changes in chemical and phase constitution of wastes, and changes in composition and properties of the reservoir material during prolonged contact with the wastes

- processes associated with the special attribute (presence of radionuclides) of liquid radioactive wastes.

Starting in the early 1960s, physical and chemical studies of the above processes were performed at the Academy of Sciences Institute of Physical Chemistry under the leadership of Academician Spitsyn and Laboratory Director Balukova.[38] In the early phase, the aim was to demonstrate the feasibility of deep injection radioactive waste disposal and to justify the requirements imposed on the wastes. Later, as the requirements on burial operations became more stringent, researchers attempted to create the proper physical and chemical composition of the wastes in reservoirs so that when wastes were disposed of as planned, their components would be immobilized.

The first task of physical and chemical research was to ensure that the wastes were compatible with the geology of the reservoir. Incompatibility of the wastes with the geologic medium would manifest itself as colmatage (plugging) of the pore space in the reservoir by suspended solids contained in the wastes or formed when the wastes contacted the rocks and subsurface waters. Incompatibility could also arise during processes that greatly complicate the technique of injection or create preconditions for emergencies, such as overheating of the reservoir, rapid gas evolution, or dissolution of formation material.

Compatibility studies involved numerous experiments to identify the conditions under which chemical reactions of wastes with rocks and waters in the formation could lead to undesirable consequences and to determine how such phenomena might be prevented. The experimental work was carried out under static and dynamic conditions, with real and simulated solutions, and with reservoir rock samples acquired from boreholes.

Dynamic filtration tests were conducted in special filtration columns simulating the pressure, temperature, and velocity of solution flow. The dynamic tests yielded limits on the concentration of suspended solids in wastes for disposal. The performance of injection wells was then analyzed, and in this way the maximum solids level (content of suspended solids leading to a decrease in flow rate when wastes are introduced into the formation) was refined. The critical loss of flow capacity was a factor-of-two (or more) reduction in the hydraulic conductivity through rock specimens.

Such a decline in the hydraulic conductivity could also come about through the formation of relatively insoluble precipitates that drop out of the wastes near the filter zone of the injection well. Such precipitates might include cations of metals either present in the wastes or leached out of the rocks by the wastes. Either effect is undesirable, and either can be prevented if the phase stability of the liquid radioactive wastes (homogeneous stability) is ensured or if the quantity of leached components is decreased by making the wastes less aggressive relative to the rocks.

If solutions containing nitric acid (0.3M) contact the reservoir rocks, which are a polymineralic mixture, there is a considerable change in chemical composition, acid-base or redox properties, and phase constitution. When rock specimens are treated with nitric acid, the cations of calcium and magnesium leach out, and the solution also picks up cations of iron and aluminum. As the nitric acid is consumed by the reaction with the rocks, the solution is gradually neutralized to pH ranging from 4 to 7. It has been found experimentally that hydroxides of such metal cations as iron, chromium, aluminum, and manganese are precipitated if pH exceeds 7.

Liquid radioactive wastes in an acidic medium must meet certain requirements relating to the continued operation of injection wells for a specified length of time. The chemical reaction of the wastes with the reservoir rocks must not destroy the rocks. To this end it has been proposed that both wastes and reservoir be pretreated, and also that liquid radioactive wastes in acidic media be injected in batches and the solutions displaced away from the wells.

On the basis of laboratory and experimental data as well as field tests, some basic requirements were formulated for liquid radioactive wastes to be injected:

- The content of suspended solids must be controlled depending on the reservoir characteristics.

- Waste composition must be controlled to prevent precipitation and gas evolution in the filter zone of the wells.

- "Threshold" concentrations must be established for waste components that are aggressive in relation to the rocks in the reservoir formation.

- The content of long-lived nuclides and those giving off the highest levels of energy must be limited to avoid overheating the reservoir.

To ensure these requirements would be met, a technology was devised for treating liquid radioactive wastes before deep injection disposal. The steps included removing suspended solids by settling or filtration, chemically pretreating the wastes, and pretreating the filter zone of the injection wells.

Requirements on wastes prepared for disposal will vary from case to case, depending on the composition and properties of the wastes, the characteristics of the reservoir, and the disposal conditions. The most general requirements are 1) limitations on the levels of finely divided solids (from 5 to 100 mg/L) and components that can form poorly soluble compounds upon relatively slight changes in the physical and chemical characteristics of the wastes, and 2) the provision that the wastes must be nearly neutral (they may be slightly acidic or slightly alkaline).

In the initial phase of radioactive waste disposal, decantates from open surficial basins were injected into reservoir formations to improve the ecological conditions in the plant areas. These were saline systems (mainly sodium nitrate) with neutral or slightly alkaline pH. No special treatment was required for these wastes.

As the main process was improved, the list of wastes destined for deep injection disposal grew much longer. The composition and properties of the wastes became more complex; pretreatment was now a necessity. The technology was developed at the Institute of Physical Chemistry of the Academy of Sciences.

Laboratory experiments, tests on real solutions, and field trials led to the view that the following steps should be included in the pretreatment:

- coagulation of easily hydrolyzed metal cations (iron, chromium, nickel), pH correction, settling

- use of compounding reagents to obtain stable solutions in which a solid phase forms at higher pH

- use of specific reagents, such as surfactants, to permit disposal of wastes varying in surface tension and viscosity.

The only components that should be removed from liquid radioactive wastes by coagulation are those that can form precipitates in the filter zone of the injection well as a result of hydrolysis or the formation of relatively insoluble compounds. The reagents (coagulants) should be chosen for each waste composition in order to bring about a safe level of precipitate-forming components.

Coagulation can be replaced by adding compounding reagents to the wastes. Chelates formed by these reagents with easily hydrolyzed waste components such as iron and chromium should be highly soluble in water over a wide pH range. This should make it possible to increase the quantity of contaminants disposed of. A number of compound-forming reagents have been thoroughly studied and can be used to treat wastes destined for deep injection disposal. For a variety of reasons, however, including availability and cost, the most practical reagent is acetic acid, whose compounding properties have been well studied.

Along with compound-forming properties, solutions containing acetate ion have "buffering" qualities; that is, when neutralized to a certain pH, corresponding to the conversion of the acid to the salt, they maintain a constant pH. Buffering is particularly important because it delays precipitation under formation conditions, so that the zone of precipitation is shifted away from the filter zone of the injection wells.

An alternative technology makes it possible to use such reagents as hydroxyethylidene-biphosphonic acid, chromotropic acid, and glyoxal in place of acetic acid when necessary. Economically, however, it is preferable to treat wastes not with costly individual compound-forming reagents but with chemical plant intermediates or wastes having the needed properties.

To this end, the Institute of Physical Chemistry of the Russian Academy of Sciences studied the possibility of using wastes from the Rubezhnoe and Erevan chemical plants. Either can be used, but implementation was difficult because of the poor solubility of the wastes in water; those from the Erevan Chemical Combine, which contained a mixture of tartaric and maleic acid isomers, also had to be refined to extract all the tartaric acid. The institute is now conducting studies to determine the possibility of using wastes from caprolactam production as a pretreatment agent for various liquid radioactive wastes.

A special method was suggested for injection of alkaline aluminate wastes; the technique is based on the amphoteric properties of aluminum, which forms soluble compounds in an alkaline medium through the generation of complex anions such as $[Al(OH)_4OH_2]^{3-}$.

Recent years have seen the wide use of pretreatment by injection of certain solutions into the filter zone of injection wells, with a transition to the injection of wastes not sufficiently compatible with the geologic medium. This approach is used on a batch basis for small quantities of wastes whose composition or physicochemical properties (viscosity, surface tension, density) differ sharply from those of the main mass of wastes. For limited quantities of wastes, the Institute of Physical Chemistry and the NIIAR (A. A. Menyailo and A. S. Ladzin) have devised a treatment and disposal process for turbid solutions having an initial suspended solids level of up to 1.0 g/L.[57] A technology has also been developed for creating "margins" in reservoirs; the purpose here is to separate wastes and subsurface waters that lack sufficient mutual compatibility.

Regardless of how wastes are categorized or pretreated, each is monitored for chemical and radiochemical composition. If necessary, physical and chemical properties, pH and redox indices, viscosity, density, and surface tension are also analyzed. Before disposal, the wastes are sampled at prescribed intervals and analyzed for an established list of components. If wastes do not comply with a particular specified parameter, they are subjected to additional pretreatment.

In the safety analysis of deep injection disposal of liquid radioactive wastes, great significance attaches to the holdup of radionuclides in the reservoir by sorption processes. In these processes, nuclides are transferred from the liquid phase (filtrate) in the pore space to the solid phase, becoming part of the mineral skeleton of the rock or precipitates on the rock surface.

In the safety analysis of deep injection disposal of liquid radioactive wastes, great significance attaches to the holdup of radionuclides in the reservoir by sorption processes.

Sorption involves physical and chemical processes including absorption, adsorption, chemosorption, and ion exchange. In addition, precipitates may form in the pore liquid (waste filtrate) and subsurface waters, also trapping coprecipitated nuclides. These processes are very hard to separate, and a generalized index called the interphase distribution coefficient is used to describe them. The interphase distribution coefficient is equal to the quantity (activity) of a nuclide contained in the solid phase (rock) to its equilibrium content in the pore fluid,

$$K_p = N/C_0,$$

where N is the activity of the nuclide in the rock, Ci/g, and C_0 is the activity in the pore fluid, Ci/cm³.

A dimensionless interphase distribution coefficient

$$K_d = K_p \rho$$

is also used, where ρ is the density of the rock, g/cm³.

The retentive properties of the rocks are characterized by the effective porosity n_{eff} and the holdup factor R:

$$n_{eff} = (N\rho + C_0 n_0)/C_0 = K_d + n_0,$$

$$R = n_{eff}/n_0 = 1 + K_d/n_0,$$

where n_0 is the active or open porosity of the rock.

The effective porosity, which accounts for the accumulation of nuclides in rocks, is physically the ratio of the total activity of a nuclide per unit rock volume (in the pore space and in the solid phase) to the specific activity in the pore fluid. The retardation factor characterizes the ratio of two effective velocities, that of water in the rocks and that of the nuclide interacting with the rocks. For example, if $K_p = 1$ cm³/g, $\rho = 2$ g/cm³, and $n_0 = 0.15$, then n_{eff} is 2.15 and $R \approx 14.3$; that is, the rate of nuclide movement due to the flow of subsurface waters is a factor of 14.3 smaller than the rate of movement of water particles and components not interacting with the rocks.

In general, it is assumed that K_p does not depend on nuclide activity (linear sorption isotherm), and that the interface distribution has an equilibrium character; the more the nuclide is sorbed, the more it is subsequently desorbed by flowing water or a solution not containing the nuclides.

Experiment has shown that the first assumption is quite acceptable for short periods of contact between wastes and rocks, because the nuclides (waste components) are present in trace amounts.

As a result, sorption capacities become saturated only when the nuclides have activities measured in curies. For example, 1 Ci of ^{90}Sr weighs 7.34 mg, while 1 Ci of ^{137}Cs weighs 11.5 mg. The activity on which migration calculations are usually based is no greater than 10^{-5} Ci/L, corresponding to 0.07 μg/L for strontium and 0.11 μg/L for cesium. At the same time, if the wastes are in prolonged contact with the rocks, they "soak in"; in this way and through diffusion processes, the area of contact between wastes and rocks increases, so that K_p increases over time.

The assumption that sorption processes are in equilibrium is an arbitrary one for convenience in nuclide migration calculations. In fact, as research has shown, only part of the previously sorbed quantity of the nuclide is desorbed; the remainder goes into solution much more slowly as a result of leaching. These assumptions lead to overestimates of the migration characteristics of nuclides in the geologic medium; this fact constitutes a "margin of safety" in calculations.

The kinetics of the physical and chemical interactions between waste components and the geologic medium can be omitted from consideration; these processes are assumed to take place instantaneously. Even so, redistribution of wastes in the pore space of rocks is a slow process that is accompanied by an increase in the effective porosity n_{eff} as the open porosity n_0 and the coefficient K_d change.

The retentive properties of the rocks depend on their type, composition, and the structure of the pore space. Other determining factors are the nuclides present, the forms they take in the solution, and the macrocomposition of the solution.

An analysis of contaminant dispersal in the geologic medium makes it possible to identify four basic types of rocks with respect to their filtration properties and their interaction with liquid wastes or contaminated waters (see Table 7). Table 8 lists the types of pore solutions (subsurface waters, waste filtrates) and describes how their macrocompositions affect the retentive properties of rocks. In Table 9, nuclides are grouped by their ability to be held up in soils and in sand-clay rocks containing fresh waters.

Studies of the retentive properties of rocks included tests of the rocks, the wastes, their interactions, and determinations of the interphase distribution coefficients for the most probable macrocompositions of wastes in various types of rocks. The greatest volume of research has been done on sand-clay rocks containing fresh hydrocarbonate-calcium waters, similar to the conditions prevailing at the disposal sites of the Siberian Chemical Combine and the Mining and Chemical Combine. Operational safety at these facilities has taken on enormous value in connection with the large volumes and high potential hazard of the wastes stored there.

Table 7. Rocks Classified With Respect to Their Retentive
Properties

Rock	Structure of Pore Space and Filtration Channels	Retentive Properties With Respect to Nuclides
Loose sedimentary rocks; sands, weakly cemented sandstone.	Uniformly distributed pores, small in size.	High
Hard igneous and metamorphic rocks: granites, porphyries, basalts, gneisses.	Network of communicating fissures varying in size.	Low
Consolidated and semi-consolidated sedimentary rocks: limestones, marls, strong sandstones.	Network of fissures and more-or-less uniformly distributed pores in blocks bounded by joints.	Low to high, depending on structure of filtration flow.
Relatively impermeable sedimentary rocks: clays, argillites, aleurolites, clay shales.	Uniformly distributed pores, quite small in size; total porosity high.	Very high

Table 8. Types of Pore Solutions Affecting the Retentive
Properties of Rocks

Solution	Macrocomposition	Predominant Processes by Which Nuclides Are Accumulated and Retained	Retentive Properties
Subsurface waters, fresh.	Salt content up to 1 g/L; sodium, calcium, magnesium; hydrogen carbonate and chloride ions.	Ion exchange; adsorption, absorption; diffusion into mineral skeleton of rocks.	High
Subsurface waters, saline.	Salt content tens or hundreds of g/L; sodium, calcium, magnesium; chloride and sulfate ions.	Adsorption, absorption; diffusion into mineral skeleton.	Low
Nonprocess wastes.	Weakly alkaline; content 1-30 g/L; hardness salts, detergents, soluble compounds.	Breaking of solution stability; formation of precipitates containing nuclides; absorption, adsorption.	High
Process wastes, alkaline.	Salt content 30-350 g/L; nitrates, soluble compounds of aluminum and silicon.	Breaking of solution stability; destruction of soluble compounds; formation of precipitates containing nuclides; absorption, adsorption.	High
Process wastes, acidic.	Salt content 250-350 g/L, pH in range of 1-3; nitrates, soluble compounds of heavy metals, compound-forming reagents.	Breaking of solution stability; destruction of soluble compounds; formation of precipitates containing nuclides; absorption, adsorption.	Low at pH less than 5-6; high after neutralization under reservoir conditions.

Table 9. Classification of Nuclides by Retainability in Soils and Sand-Clay Rocks[39]

Nuclide	Form of Nuclide in Subsurface Waters	K_p, cm³/g
Tritium	In water molecules	0
Technetium, iodine, carbon, phosphorus, sulfur, chromium	Anionic	0 to 10
Strontium, radium, calcium, and other alkali-earth metals	Cationic; poorly soluble forms of sulfates and carbonates	5 to 50
Cerium, promethium, manganese, iron, cobalt, zinc, zirconium, neptunium, and other rare-earth and transition metals	Hydroxide complex forms	10 to 100
Cesium	Stable cationic	100 to 1000
Thorium, plutonium, americium	Polymeric	100 to 1000
Note: The retainability of a nuclide will vary with the composition of macrocomponents in the pore solution.		

Experiments were done on rock specimens acquired from the reservoir horizons during well drilling. Analysis by a variety of methods (x-ray radiography, x-ray diffraction, infrared spectroscopy, chemical analysis) showed that the most typical minerals composing the rocks are quartz (70 to 80%); feldspars of the orthoclase, microcline, or plagioclase type (10 to 15%); mica- and hydromica-group minerals (muscovite, phlogopite, vermiculite); and clay minerals of the kaolinite (not less than 3 to 5%) and montmorillonite groups (8 to 10%). Some specimens included carbonate minerals and organic matter. The exchange complex has the cations Ca^{+2}, Mg^{+2}, K^+, Na^+, and H^+.

The sorption capacity of sandstones is largely determined by the amount of clay minerals, which are known to be good natural sorbents.[40] The mechanism of sorption depends on the character of the crystal lattices of the minerals. Particles of kaolinite have a rigid lattice, and ions in the solution cannot gain access to the spaces between sheets. The basal surfaces are electrically neutral, so that exchange reactions occur only where the lattice is deformed (sheared). The cation-exchange capacity is 3 to 5 mg-equiv/100 g.

Hydromicas also have a rigid crystal lattice, but the basal surface bears a sizable negative charge, so that cation exchange takes place not only at deformations but over the entire basal surface. The capacity can range up to 10 to 40 mg-equiv/100 g. In both kaolinites and hydromicas, ions are exchanged only at the external surface of a particle.

In clay minerals of the montmorillonite group, the grain surface may be a torn region of the crystal lattice and thus consists of atoms having uncompensated electrovalent bonds. Ion exchange can occur not just at the external surface but also at lattice distortions, and so the capacity is up to 150 mg-equiv/100 g.

The sorption capacity of feldspars is governed by the presence of exchange cations—sodium, calcium, and protons of hydroxyl groups—at the surface. In quartz, the sorption capacity is due to adsorption phenomena and does not exceed 0.5 mg-equiv/100 g. The sorption capacity of a polymineralic material is a weighted average based on the composition; it depends on the ratio of sand, silt, and clay fractions. Sorption in such a case is the net result of several parallel processes.

Table 10. Sorption Properties of Sand-Clay Rocks

Specimen Sorption	Content of Clay Fraction, %	Capacity, mg-equiv/100 g
Sandstone, clayey	18-24	3.0-3.5
Sandstone	10-20	2.5-3.2
Sandstone, slightly clayey	8-15	2.6-2.8
Sand, fine	7-8.5	2.5-2.7
Sand, medium	6-7	1.8-2.0
Sand, coarse	5-6	1.0-1.6
Sand, washed	3-5	1.0-1.5

The sorption capacity of sand-clay rock specimens typical of reservoirs is determined from the ^{90}Sr saturation against a sodium nitrate background. Table 10 presents results from these determinations.

A delay in the migration of waste nuclides depends not only on the sorption capacity of the rocks, but also quite strongly on the nature of the processes in the waste/formation water system, the form of the nuclides, the tendency of the wastes to form precipitates when mixed with subsurface waters, and the properties of these precipitates. For these reasons, the retentive properties of rocks were studied with real wastes or model solutions under conditions fairly close to the formation conditions.

The distribution of nuclides was investigated under static and dynamic conditions. In the static tests, specimens of disintegrated loose rocks were shaken together with the solution, the mixture was allowed to separate, and the nuclide contents were measured in the solid and liquid phases. In dynamic tests, the solution was filtered through disintegrated or intact rock specimens, and the nuclide contents were determined in the effluent. The rocks were first saturated with water similar in composition to the subsurface waters.

The static and dynamic tests yielded quite different results. The processes under static conditions were similar to those when solutions are resident for extended times in the rocks, moving at low velocities, so that the rocks become "soaked" with the solutions. The dynamic behavior resembled the processes that take place when solutions move rapidly through rocks, for example near the injection wells and in the early years of disposal site operations. The desorption of nuclides from the rocks by various solutions was also investigated.

Table 11 presents interphase distribution coefficients K_p determined for the principle waste fission-product nuclides at various salt contents (sodium nitrate) and acidities. Plutonium-239 was also included since it is present at unrecoverable trace levels, close to the permissible DKb values, in wastes. This table shows that the retentive properties of sand-clay rocks tend to increase with increasing pH and decreasing salt content.

Table 11. Interphase Distribution Coefficients K_p (cm^3/g) of Nuclides in Sand-Clay Rocks

Nuclide	Pore Solution			
	pH = 2-3 μ = 1.0	pH = 4-5 μ = 1.0	pH ~ 8 μ = 0.3-0.5	pH ~ 8 μ = 0.1
^{90}Sr	1.5-5.5	10-35	7-10	20-30
^{106}Ru	0.5-1.5	7.5-15	2-2.5	4.6-7.5
^{137}Cs	1.5-3.0	10-20	8-15	20-50
^{144}Ce	1.0-1.5	40-100	5-10	9.5-19
^{239}Pu	1.2-1.6	50-120	5-12	15-35

Note: μ is the ionic strength, defined as the salt content. K_p values in the pH 4-5 column were obtained for conditions of hydrolysis of macrocomponents.

Interphase distribution coefficients and nuclide accumulation figures for slightly saline and slightly alkaline nonprocess wastes appear in Table 12. The wastes are very similar in composition to formation waters. The table also gives retardation factors and extents of desorption by water for sand-clay rocks containing fresh waters.

The liquid phase of such wastes is unstable; under formation conditions, it drops a precipitate, chiefly hydrated aluminum oxides, promoting the elimination from the solution of all colloidal

Table 12. Retentive Properties of Sand-Clay Rocks for Weakly Alkaline Nonprocess Wastes From Radiochemical Plants

Nuclide	C_0, Multiple of C_{DKb}	K_p, cm³/g	Holdup Factor R	Desorption, %
^{90}Sr	$5 \cdot 10^3$	35-37	200-300	5.5-9.0
^{106}Ru	$9 \cdot 10^2$	7-14	45-85	1.4-4.0
^{137}Cs	$2.7 \cdot 10^2$	60-90	350-450	1.5-2.0
^{144}Ce	$8 \cdot 10$		95-115	1.5-4.0
^{239}Pu	10	110-140	500-625	3-6

Note: C_0 is the content of the nuclide in the wastes.

and pseudocolloidal forms of radionuclides in the solid phase. Aluminum hydroxide is a good sorbent and performs this role in the waste/formation water system for trace levels of nuclides. The interphase distribution is described by large K_p, especially for elements forming relatively insoluble compounds with the anion of the precipitate (cerium, plutonium). Strontium and cesium are absorbed from the solution by an ionic adsorption mechanism on the precipitate surface, since in alkaline environments (pH ~ 9) hydrated alumina is a cation exchanger.

Ruthenium is captured in the solid phase to a lesser extent than the other nuclides. In this medium, ruthenium takes the form of large hydrolyzed ions bearing a negative charge. But even so, the rate of nuclide migration is more than a factor of 10 below the velocity of water and of components not reacting with the rocks.

Fairly strong attachment of nuclides to rocks has been noted. When pure water flows through the sorption zone, no more than 10% of the nuclides in the solid phase are transferred to the liquid phase; the remainder may enter the water by leaching—that is, very slowly and over a long span of time. The resulting contents of nuclides in subsurface waters will be well below the DKb values.

Process wastes from radiochemical plants are more complex systems with high-salt contents and high activities. These wastes are classed in two major types, alkaline and acidic. The overall behavior of alkaline wastes in the pore space of the reservoir is similar to the behavior of nonprocess wastes. The K_p is lower, however, because of the high-salt background (see Table 13). The quantity of cations (cesium, strontium) absorbed by the rocks themselves will depend on the cation-exchange capacity. In alkaline media, the exchange capacity increases when protons of the hydroxyl groups °Si—OH and °Al—OH located on the lateral faces of the clay mineral crystals take part in exchange reactions.[41] For the hydrolyzed forms of nuclides (ruthenium, plutonium),

Table 13. Retentive Properties of Sand-Clay Rocks for Alkaline Process Wastes From Radiochemical Plants

Nuclide	C_0, Multiple of C_{DKb}	K_p, cm^3/g	Holdup Factor R	Desorption, %
^{90}Sr	$2.5 \cdot 10^8$	5.5-7.0	25-30	16-31
^{106}Ru	$6.8 \cdot 10^6$	6.0-10.5	30-50	10-30
^{137}Cs	$6 \cdot 10^6$	4.5-6.5	20-30	10-13
^{239}Pu	$1.4 \cdot 1000$	40-95	200-460	10-15

ion exchange may not occur at all, but other mechanisms are possible: precipitation of hydrolysis products or physical trapping of colloidal particles.

According to M. K. Savushkina of the Institute of Physical Chemistry, the dioxides of silicon and aluminum undergo a transformation when alkaline wastes interact with sand-clay reservoir rocks; these species are later converted to zeolite-like minerals. Their sorption properties are substantially greater than those of the rocks.

Acidic process wastes destined for disposal contain, as macrocomponents besides sodium nitrate salts, soluble complexes including heavy metal (corrosion product) ions: iron, chromium, manganese, and nickel. If such wastes are injected without pretreatment, the interaction of the wastes with the rocks leads to a drastic decrease in acidity; weakly soluble hydroxides precipitate, also trapping coprecipitating nuclides. A high degree of accumulation in the rocks may lead to overheating of the formation, creating the preconditions for accidents. Two approaches can be taken to stabilize the solutions and limit the accumulation of nuclides: lower the acidity to render the wastes less aggressive, and treat the wastes with complex-forming reagents such as acetic acid to complex the easily hydrolyzed cations. The acidity can be decreased to pH 2 to 3, and the hydrolyzing heavy metal cations can be stabilized through the formation of acetate complexes. As a result, the accumulation of nuclides in rocks and their retentive properties are diminished. Table 14 characterizes the retentive qualities of sand-clay rocks for such systems.

These studies showed that acetate complexes have limited stability, breaking down in the reservoir formation through the action of temperature and radiation, so that the acidity decreases. As a result of these phenomena, hydrolysis sets in and hydroxides (such as those of iron and aluminum) precipitate, causing K_p to rise to values close to those shown in the pH 4-5 column of Table 11. The K_p for transuranic nuclides increases by a factor of 10 or more.

This phenomenon has been used to develop pretreatments and batchwise injection of acidic process wastes. The addition of a

*The good
retentive
properties of
sand rocks
for waste
nuclides found
in laboratory
trials have
been confirmed
by observations
at an actual
disposal site.*

Table 14. Retentive Properties of Sand-Clay Rocks for Acidic Process Wastes From Radiochemical Plants

Nuclide	C_0, Multiple of C_{DKb}	K_p, cm³/g	Holdup Factor R	Desorption, %
^{90}Sr	$5 \cdot 10^9$	1.2-1.3	6-9	21-22
^{106}Ru	$7 \cdot 10^7$	1.3-1.6	6-7	0.5-1.2
^{137}Cs	$2 \cdot 10^7$	2.7-2.8	13	0.7-1.3
^{144}Ce	$6.0 \cdot 10^8$	1.0-1.6	4-8	12-19
^{237}Np	10^2	1.8-2.3	9-11	8-40
^{239}Pu	10^3	1.4-6	13-30	0.5-3.0
^{241}Am	10^4	1.1-1.2	5-6	18-20

complexing reagent makes it possible to convert up to 85 to 90% of nuclides into solid form after injection, or to partially solidify them directly in the formation, though outside the filter zone.

The good retentive properties of sand rocks for waste nuclides found in laboratory trials have been confirmed by observations of waste migration in a collector layer at an actual disposal site. For example, when nonprocess wastes from the Siberian Chemical Combine and the Mining and Chemical Combine were injected, observations in instrumented wells 250 to 400 m away from the injection wells revealed only nonradioactive components (anions of the salts). The effective thickness of the collector layer was 20 to 30 m, and open porosity was 0.1 to 0.2; 1 to 2 million m³ of wastes was injected. At the same time, the gamma field measured in the injection wells in the collector layer intervals was three orders of magnitude higher than if no nuclides had been accumulated in the rocks. Calculations showed that the K_p corresponding to the observed accumulation for gamma-emitting waste components was on the order of 1 to 3 cm³/g. Temperature measurements in an injection well where acidic process wastes were disposed of showed a moderate accumulation of nuclides in the rocks; the corresponding K_p would have been 0.5 to 1 cm³/g.

The most forceful confirmation of the retentive properties of sand-clay rocks, however, came from an injection test with intermediate-level wastes, a decantate obtained when pool B-2 of the Siberian Chemical Combine was taken out of service. The wastes were injected at 1900 to 2000 m³/d into a reservoir in horizon II, lying at a depth interval of 326 to 349 m. The subsurface waters were sampled in an instrumented well 55 m away from the injection well. After 50,000 m³ of wastes had been injected, the samples exhibited levels of nitrate and chloride ions (species that do not interact with the rocks) equal to the levels in the wastes, but strontium and ruthenium were three orders of magnitude below the wastes, and cesium was four orders of magnitude lower.

Figure 7 is a diagram with plots of the test results in the form of "output" curves for certain components.

The good retentive properties of porous rocks for nuclides have also been remarked on by other workers. According to E. V. Sobotovich and Yu. A. Ol'khovik, the sand rocks in the embankment of the holdup pond at the Chernobyl nuclear plant exhibit K_p values of 5 to 6 cm^3/g for ^{90}Sr and 150 to 300 cm^3/g for ^{137}Cs.[42] U.S. studies of the retentive qualities of various rocks, performed for the U.S. Environmental Protection Agency,[43] showed sandstones containing fresh water to have K_p values of 2.5 cm^3/g for strontium, 200 cm^3/g for cesium, 2000 cm^3/g for plutonium, and 20,000 cm^3/g for americium. For clays, the K_p values for plutonium and americium were several times greater. A large number of such determinations are generalized in a publication already cited.[39]

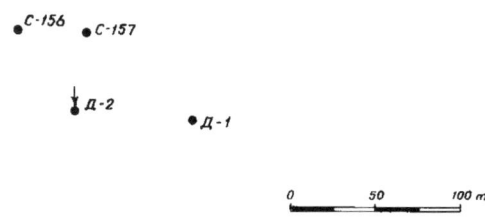

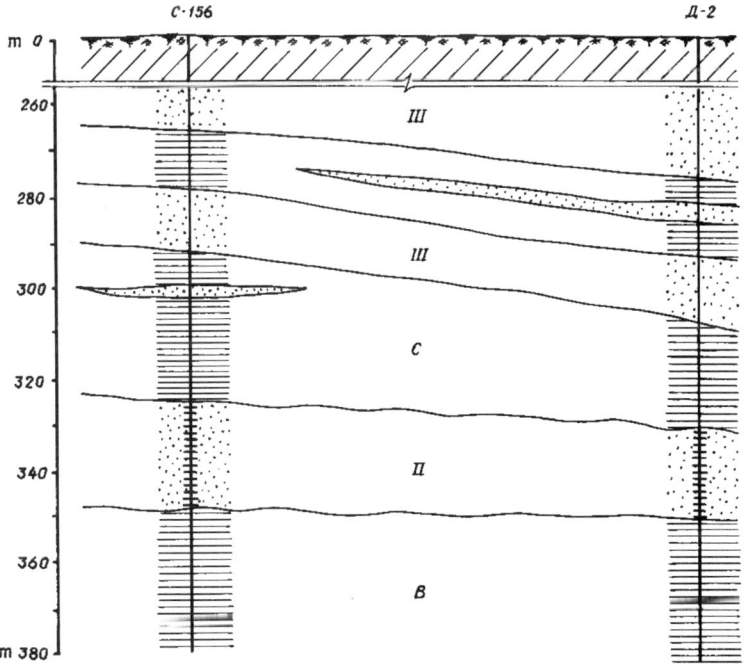

Figure 7a. *Map Showing Well Locations and Geologic Section for Well S-156*

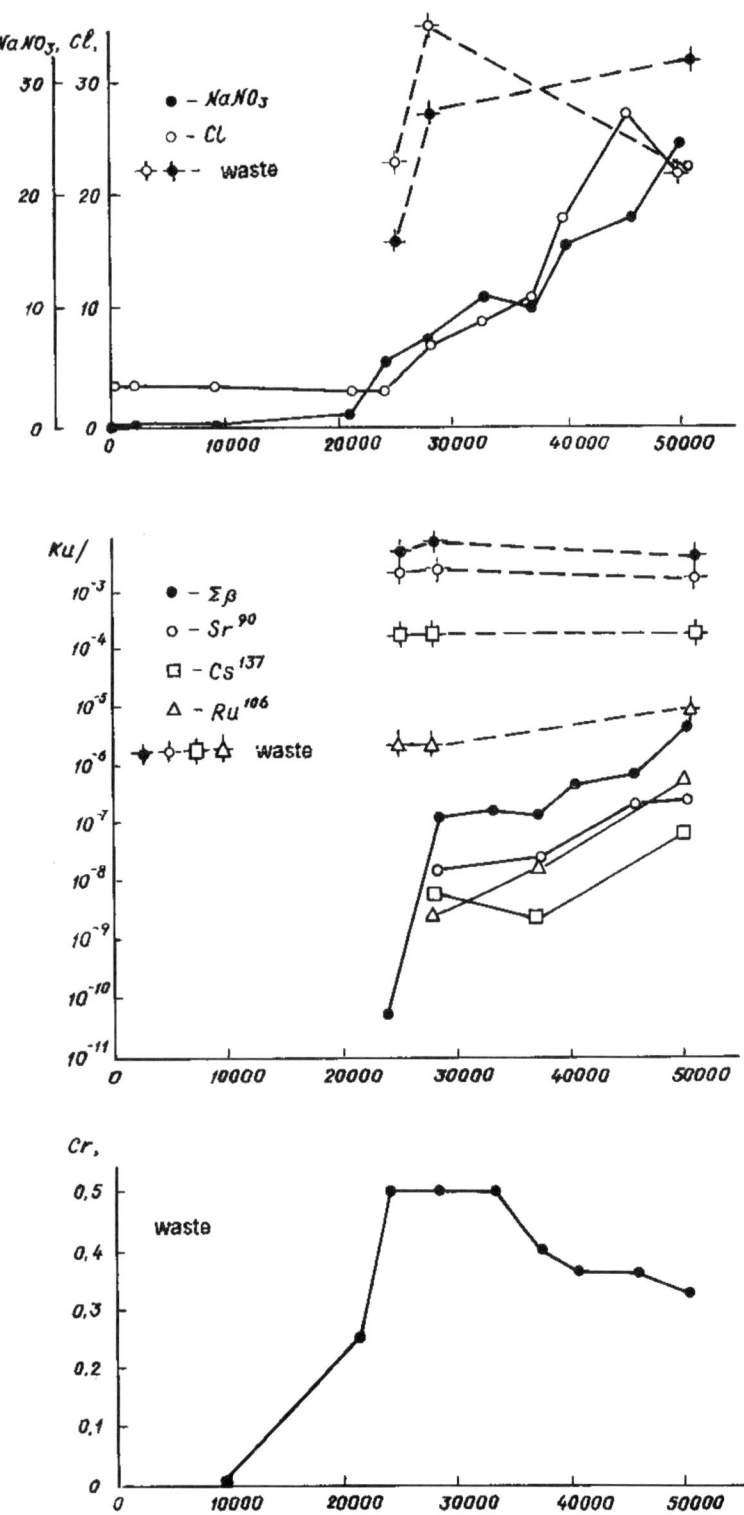

Figure 7b. *Fluid Composition in Well S-156 as a Function of Volume Injected*

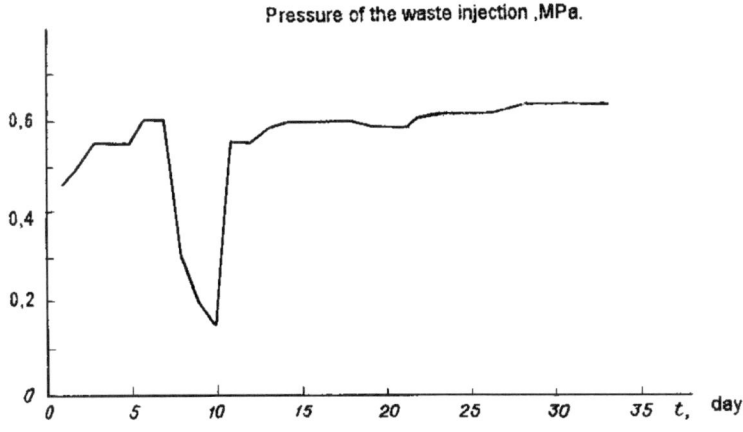

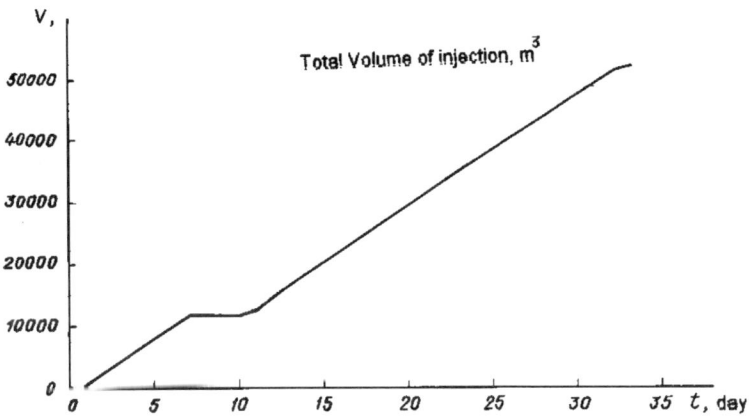

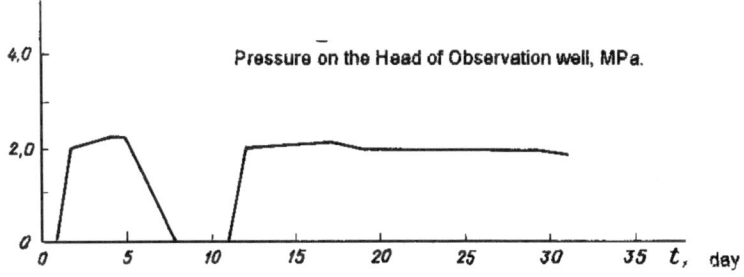

Figure 7c. *Well S-156 Conditions During Injection*

3.4 Models of Disposal Processes

Forecasting the consequences of the deep injection disposal of liquid radioactive wastes is required when using this method of waste handling. Calculations of reservoir pressure conditions and migrations of waste components are carried out to determine the area of influence of the waste burial, to delimit sanitary protection areas and subsurface exclusion zones, and to select optimal injection patterns and monitoring setups. The safety analysis includes predictions of other processes involved in deep injection disposal.

Models of the geologic medium and processes used to forecast the consequences of disposal are justified on the basis of geological exploration, filtration tests, laboratory trials, and experience with similar practical problems. A model is a conceptualization of objects or phenomena, qualitative in its early phase and complemented by quantitative relations as research and development results are obtained. Such relations are usually expressed in a system of equations and sets of parameters and characteristics, including data in the form of maps, diagrams, sections, and tables.

In the initial phase of work on deep injection disposal of liquid radioactive wastes, theories of petroleum hydrodynamics devised by V. N. Shchelkachev, P. Ya. Polubarinova-Kochina, and others were employed.[44, 45, 46, 47] In this way, needed calculations of pressure fields and reservoir filling by injected solutions could be performed given certain assumptions.

Predictive calculations, done for simplified conditions, featured a large margin of reliability, but this did not stand in the way of acquiring needed starting data in order to justify and design burial systems or demonstrate the safety of this waste handling method. Further work was focused on perfecting the theory of radioactive waste disposal, and models of geologic conditions and processes. Also, and this must be stressed, methods of investigating mathematical models and solving predictive problems were carried out subsequently, as disposal sites were designed, constructed, and operated; injection regimes were optimized; and disposal safety was confirmed. Computer techniques of numerical simulation found wide use in the 1970s and 1980s and led to the solution of many problems in models emulating the conditions of liquid radioactive waste disposal.

Major roles in developing the theory and practice of predictive calculation were played by N. N. Verigin, B. S. Sherzhukov, V. S. Sarkisyan, and F. M. Bochever, staffers at the All-Union Scientific Research Institute of Water Supply, Sewerage, Hydrotechnical Structures, and Engineering Hydrogeology (VNIIVodgeo), who worked on hydrodynamics and mass transport with physical and chemical interactions and hydrodynamic dispersion. Researcher V. M. Gol'dberg of the All-Union Scientific Research Institute of

Hydrogeology and Engineering Geology studied the flow of solutions in the clays of confining formations and the flow structure of injected solutions. Research by V. M. Shestakov (Moscow State University) and V. A. Mironenko (Leningrad Mining Institute [now St. Petersburg Mining Institute]) in the protection of subsurface waters and the consequences of contamination was also useful.

The VNIPIpromtechnologii did a great deal of work on the performance of injection wells (P. P. Kostin) and on thermal fields in the injection of liquid radioactive wastes (F. P. Yudin, L. I. Vladimirov, G. A. Okun'kov). This institute developed detailed numerical techniques for the study of theoretical models giving the closest possible fit to real conditions (Okun'kov, M. A. Sinel'nikov, and N. M. Burmistrova). The results of predictive calculations were validated in the operation of disposal ranges. The following section will describe in general terms the models and methods in use as of late 1992.

Processes that accompany the deep injection disposal of liquid radioactive wastes can be classified in four main groups by the generality of the models employed:

- The variation of the peizometric surface (pressure regime) of subsurface waters in a reservoir into which wastes are injected.

- The filling of the pore space and the distribution of wastes in the reservoir and confining structures; mixing of wastes with subsurface waters; physical and chemical phenomena in the system comprising wastes, rocks, and formation waters.

- The variation in the geostatic condition of the geologic medium.

- The change in temperature of the geologic medium.

These processes take place in a geologic space whose characteristics largely control the parameters of the processes. One of the first steps in prediction is, therefore, to justify the models of the geology and, in particular, the models of the spatial distribution of the hydrogeologic parameters. These parameters are known at discrete points or in regions, that is, where wells have been drilled and filtration tests have been run. Forecasting requires the use of intermediate values, which for some parameters can be calculated by a local approximation method.

The variation of the pressure regime in the reservoir where wastes are injected is the first key indicator of waste entry into the pore space and the displacement of the subsurface waters. The change in pressure (head) manifests itself in rising piezometric levels of subsurface waters in observation wells (decreasing depth to the waters) or rising pressure at the wellhead if the level of the subsurface waters is above the Earth's surface, and the wellhead is sealed.

A "repressuring dome" forms in the reservoir; its central part is located near the injection well. The pressure declines roughly as the logarithm of the distance from the well, while the parameters of the repressuring dome are determined by the time and volume of injection; the geologic conditions and characteristics of the reservoir formation; and the design, number, and layout of the wells.

If the pressure in the reservoir is relieved, for example by pumping formation water out of wells distant from the injection wells, a cone of depression forms, and the developed pressures become lower than in the no-relief case. A hydrodynamic perturbation in the reservoir propagates significantly faster than the wastes themselves, so that the pressure field makes it possible to estimate the structure of the waste flow field long before the wastes appear. Figure 8 shows a repressuring dome and waste dispersal in a reservoir under arbitrarily chosen conditions.

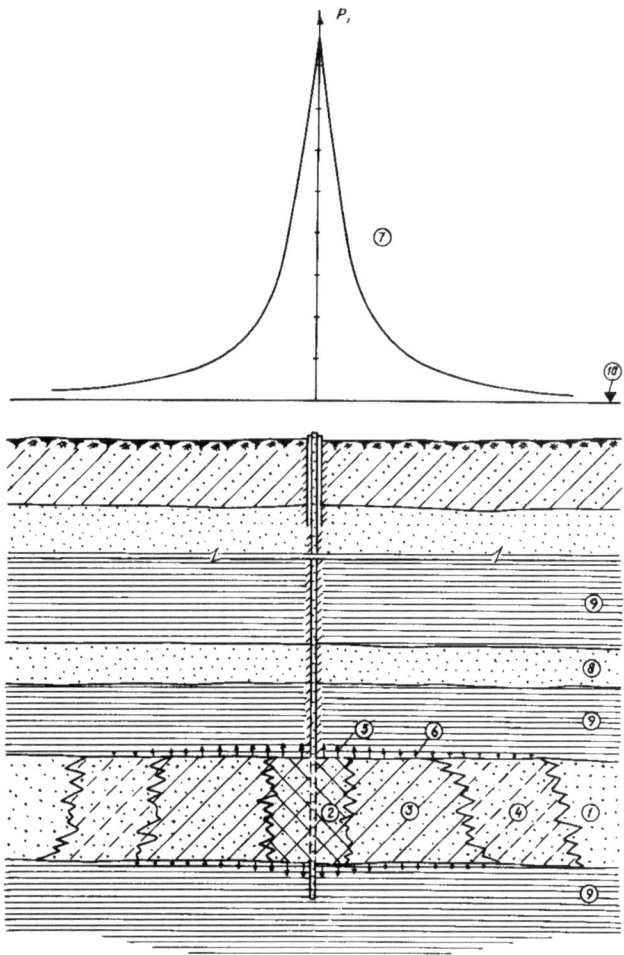

Figure 8. *Repressurizing Dome; Dispersal of Wastes in Reservoir Formation. 1 - reservoir; 2 - zone of nuclide accumulation; 3 - filtrate of wastes; 4 - zone of dispersion; 5 and 6 - solutions entering confining formations by flow (filtration) and diffusion; 7 - formation pressure curve; 8 - buffer horizon; 9 - relatively impermeable horizons; 10 - natural formation pressure.*

Natural pressures and the pressure fields set up by waste injection determine the direction and velocity of liquid flow in the porous medium of the reservoir; the direction and speed also depend on the flow characteristics of the reservoir, the boundary conditions, the layout of the wells, and other factors. Pressure fields are described by the well known filtration equation, for which analytic solutions are known for extremely simple cases.[47-53] Analog simulation methods have been widely used in the justification and design of disposal ranges (electrohydrodynamic analogy method). Two-dimensional filtration equations have been solved in this way for realistic reservoir geometries and injection patterns.[54]

The finite-difference method has been used more recently to solve filtration equations. The system of linear equations obtained in the finite-difference approximation is solved by componentwise over-relaxation. Such a flow model makes it possible to simulate two flow regimes: with and without sources. If sources are active, such as injection and relief wells or water intakes, their flow rates are approximated by a step function, and the entire flow regime is considered a succession of steady states. The natural regime of subsurface waters is simulated without sources.

When a concrete model of an aquifer is developed, there may be no accurate information about the pressures at some points. For a number of reasons, a directly written system of equations cannot be employed in such a case. The following variant method has been proposed for constructing a flow model. The aquifer is outlined in plan by a boundary passing through marginal points where experimental pressure values are known. Intermediate values of the pressure on this boundary are calculated by a local approximation method, and the flow equation is solved for the interior. The procedure yields a distribution of pressure fields for various periods of time, as well as injection and relief schemes that can be used in constructing flow lines of the formation fluid; these schemes form the basis for calculating waste dispersal.

Pressure fields based on geological exploration data are commonly somewhat different from those found during waste injection. The reasons are the inevitable uncertainties in starting data, together with interpolation and averaging errors. Experience has shown that defective injection wells, relating to the way in which the formation is tapped, sometimes lead to a significant loss of pressure in the screen and filter zone of the well. For this reason, water is injected for a considerable time after the completion of operational wells. The piezometric surface of the subsurface waters is monitored in observation wells during this period, and data from these wells make it possible, by solving inverse problems, to correct the reservoir models and refine the predictive calculations of pressure and flow nets. If necessary, the known flow equation can be modified to include interactions between horizons, the macroscopic nonuniformity of the formations, and other factors.

After waste injection has been completed, the pressures in the central part of the repressuring dome decline through redistribution of the head over large areas. The reduction in excess pressure in injection wells below the Earth's surface is a function of the formation flow characteristics and the volume and duration of injection (from days to months). If wastes with low-salt levels are injected into formations with mineralized waters, the wellhead pressure remains elevated even after waste injection has been concluded.

The dispersal and migration of wastes and their components in the reservoir are governed by the induced and natural pressure fields, the volumes and conditions of injection, the capacities of the formation, its nonuniformity, and physical and chemical reactions in the system comprising wastes, rocks, and formation waters. Because of the pressure gradient that develops, the wastes move from the borehole through the well screen and into the reservoir, where they fill the pore space, displacing and mixing (partially) with subsurface waters.

As a result, a "pool" of wastes appears in the reservoir. A dispersion (mixing) zone of wastes and formation waters is located at the margin of the pool. Physical and chemical processes in the waste-rock-water system cause the waste composition in the pore space to change as waste components are transferred to the solid phase (as sorbate on rocks and precipitates and as the actual precipitates). The acidity of the waste filtrate changes, with the pH approaching the natural value; this is another factor aiding precipitation (see Section 3.3).

The greatest accumulation of nuclides in the reservoir rocks usually occurs close to the injection wells. The components in the marginal zones of the pool are mainly those that do not interact with the rocks, along with the water that is present in the wastes. Waste components diffuse into the rock matrix of the reservoir and the confining structures that underlie and overlie it. After waste injection is stopped, waste components are displaced by the natural movement of subsurface waters (see Figure 8).

A two-dimensional plan model of mass transport was used to describe the dispersal and migration of radioactive waste components in reservoirs. The model includes convective transport, hydraulic dispersion, and radioactive decay. The interaction of nuclides with the geologic medium is handled by introducing an effective (apparent) porosity, which depends on the interphase distribution coefficient of Section 3.3.[47, 51, 55] Similar methods are used to model the nonequilibrium nature of sorption processes (partially irreversible sorption).

The mass-transport equations are solved by finite differences, but here a new problem arises. The main error in the numerical solution comes from the finite-difference approximation of the convective terms in the mass-transport equation ("numerical dispersion"). At low Peclet numbers, traditional approximations of the convective terms can be employed. But at large Peclet numbers, when contaminants are highly toxic and levels on the order of 10^{-5} to 10^{-10} must be tracked, the effect of numerical dispersion can be quite large. The principal way of improving the accuracy of the solution in such cases is to break the spatial mesh into smaller cells, which leads to a drastic increase in the demands made on computer resources. One proposed method of cutting the numerical dispersion without significantly increasing machine requirements is to modify the finite-difference method by treating convection and dispersion separately.

The equation of dispersion (with no convective terms) is solved in a two-dimensional difference mesh (basic mesh); the two-dimensional equation of convective transport reduces to a one-dimensional equation within each cell of the basic mesh. This reduction in dimension permits a significant decrease in the spatial discretization interval of the convective term, which improves the accuracy of solution without greatly increasing the demands on computer memory and machine time. The parameters in the equations are determined from geological exploration data, field tests, and laboratory trials.

This method of calculating waste dispersal and the migration of waste components was used to forecast the processes and consequences of the deep injection disposal of liquid radioactive wastes from the Siberian Chemical Combine and the Mining and Chemical Combine. Some of the results are presented in Chapter 5, which is devoted to the analysis and assessment of the safety of operational injection areas. The finite-difference solution is very unwieldy, requiring a considerable amount of computer time. Simplified expressions given in the cited literature can be used for rough calculations.

The difference in density between injected wastes and subsurface waters leads to the formation of a plug-flow (displacement) front that is inclined in the vertical plane. The net effect is to increase the travel distance of waste dispersal compared to that of the plug-flow. Computational formulas are known for estimating the width of the inclined boundary projected onto a horizontal plane, but these do not include dispersion phenomena.

The projected width of the inclined boundary increases with increasing difference in specific gravity between the contacting wastes and the formation waters. Dispersion phenomena decrease this difference in the zone where wastes and water mix, since the density gradient becomes smaller upon mixing. The inverse proportionality between the width of the density-related inclined boundary and the width of the dispersion zone has also been pointed out.[55]

Okun'kov at the All-Union Scientific Research and Design Institute of Industrial Technology wrote a program to predict waste migration in a synclinal environment with a density difference (density convection), hydraulic dispersion, and viscosity. (Results from these calculations are presented in Section 5.2.) A comparison of results with and without density effects shows that this factor has an impact for long time spans in the central portion of the waste contour. This is the region where the density difference relative to subsurface waters is greatest. The comparison applies when the central region is located in a synclinal zone and the reservoir levels are minimal.

An important factor diminishing the effect of the density difference between wastes and subsurface waters is the bedded structure of the reservoir. Reservoir formations can be visualized in cross section as a succession of beds varying in permeability and separated by beds of low permeability. The permeable layers are typically thin, 5 to 15 m, so that the projected width of the inclined boundary is also less than if the total thickness of all beds were taken.

The bedded character or vertical nonuniformity of reservoir formations has a strong influence on waste dispersal, especially when beds exist with much higher flow properties than the section average. If geological exploration reveals such a situation, waste movement must be calculated bed by bed; this means solving two-dimensional profile problems. At the same time, observations in three radioactive waste injection sites have shown that vertical nonuniformity becomes apparent at fairly small distances from the injection wells (tens of meters to a few hundred meters). The waste front in large rock masses evens out over the beds with increasing distance from the injection system. This happens for several reasons: facies variation and lithologic replacement of highly permeable beds; wedging out of relatively impermeable beds separating them, giving rise to flows between beds; and variation in the injectivity profile as a result of colmatage.

An exception occurs in anomalously permeable beds that differ from the main reservoir in lithology and pore structure. An example is the thin bed of vuggy calcareous sandstones lying near the roof of the terrigenous bench in permeable zone III of the NIIAR pilot injection area (Section 5.3). Such beds can be identified by special studies included in geologic exploration.

The predictive calculations account for horizontal nonuniformity of the flow properties by using a permeability distribution in the reservoir region under consideration. The distribution is based on geologic exploration results.

Calculations done in the early 1960s, when candidate injection areas were being characterized and designed, usually did not account for nuclide retention by rocks. Determining the position of the "liquid" or aqueous phase of the wastes gave a result with a large margin of safety. Later, effective porosity values based on the open porosity and the K_p were used (see Section 3.3). It was thought, however, that sorption processes were reversible, in other words, that all the nuclides sorbed would also be desorbed by the flow of the subsurface waters. In fact, the lesser part of the nuclides is desorbed. Partially irreversible sorption decreases the contents of the nuclides in the subsurface waters to values below the DKb and causes the nuclides to be incorporated into the rocks near the injection wells.

The fraction of pores filled by wastes also shows an increase over time. Diffusion carries waste components into dead-end and subcapillary pores as well, so that the porosity of the reservoir rocks, which governs their capacity, increases until it approaches the total. The interphase distribution coefficients for the nuclides also increase. All these factors tend to decrease waste dispersal in the reservoir.

A rise in pressure in the reservoir will create the conditions for wastes to move into relatively impermeable clay confining rocks lying below or above the reservoir. While the confining bed has low flow properties, they are not zero. According to studies by Gol'dberg, solutions will flow in clays once the pressure gradient passes a threshold value called the initial filtration gradient.[24] Such conditions occur close to an injection well in the central part of the repressuring dome.

The low filtration factors of clay confining beds also imply low velocities of flow through clayey rocks. These rocks also have high retentive properties, so that waste components do not reach the shallower buffer horizon. The penetration of wastes into clays will occur only during the injection period when overpressures are present.

If low-salt content wastes are injected into reservoirs containing highly mineralized waters, an excess pressure (pressure greater than the natural pressure in the reservoir) will be observed even after injection is terminated. Flow in the clays, however, is much reduced by the considerable decline in permeability when the saline pore fluid is replaced by fresh water.[24]

The penetration of wastes (liquid or aqueous phase) into underlying and overlying clays, where the initial filtration gradient is exceeded, can be calculated by a finite-difference method. More often, however, an approximate solution is employed.

Appreciable flows between the reservoir and shallower horizons occur when filtration windows (lithologic or tectonic) exist in the confining beds. It is thought that no such windows are found in disposal regions, in accordance with geologic exploration results. Otherwise, the decision to proceed with disposal is not made.

Nevertheless, estimates are necessary to analyze hypothetical complications and accident situations. A leak of liquid through a window could begin before wastes arrive in the region (unless the window coincides with the injection contour). Such a flow can be detected by monitoring the initial stage of deep injection disposal, so that steps can be taken to change the injection conditions (up to suspension). There are ways to identify and locate leakage zones and determine the rate of flow. In some cases, a well can act as a window if the drill string/casing annulus and the casing/borehole annulus are not isolated.

The movement of waste components into beds above and below the reservoir due to concentration gradients can generally be described by Fickian diffusion. Calculations show that diffusion plays a minor role in the dispersal of nuclides out of the reservoir. For example, if ^{90}Sr is present at a level of 1 Ci/L, it will diffuse no farther than 10 m into an overlying confining bed in the time required for the nuclide to decay to safe levels.

The development of a repressuring dome in the reservoir redistributes stresses in the rocks, which in turn leads to deformations. Under steady-state injection conditions, these changes are quite small, so that there is no perceptible displacement of the strata. For example, if wastes are injected into a reservoir at a pressure of 2.0 MPa at a depth of 400 m when the static level of subsurface waters is 20 m, and if the waste density is close to that of the subsurface waters, then the increase in the natural formation pressure in the reservoir would be no more than 15 to 20% in the immediate vicinity of the injection well. The relative increase in formation pressure falls off as the logarithm of the distance from the well. If low-salt content wastes are injected into deep reservoirs containing highly mineralized waters, the change in the natural formation pressure will be 0.5 to 1 order of magnitude less.

There are equations that describe how the stress-strain state affects the consolidation of rocks lying above permeable horizons from which water is withdrawn or into which solutions are injected.[56] These equations can be used to calculate deformations and displacements. Solving them leads to approximate estimates of the possible geodynamic behavior.

The calculations show that the change in position of the surface due to changing formation pressure is measured in fractions of a millimeter. These displacements fall off in intervals more than a few hundred meters above the reservoir. As a result, surface movements due to waste injection can usually be detected only when shallow horizons are used.

The surface reaction to the development of a repressuring dome is significantly delayed and can be estimated with known formulas. The lag time, which depends on the depth of the reservoir and the pressure developed, ranges from years to centuries.

In sedimentary complexes suitable for the deep injection disposal of liquid radioactive wastes, there are generally horizons of rocks having more-or-less elastic qualities not considered in geodynamic calculations. A change in the stresses in the geologic medium, therefore, has minimal—in practice, imperceptible—effects. However, waste injection pressures are regulated and must not exceed the hydraulic fracturing pressure. Higher pressures would alter the structure of the reservoir and the position of its rocks.

Surface subsidence resulting from rapid lowering of the water table has been observed. In such cases, the volumes of water that are produced are incomparably greater than the volumes of injected wastes, with lowering of the levels and drying of the upper zones of shallow aquifers, as well as associated changes in the engineering-geologic properties of the rocks. Geodynamic phenomena and induced seismicity occur when high-pressure waters are injected directly into tectonic fault zones where the contacting blocks are made of hard rocks in a state of stress, that is, in seismically active areas and active fault zones where the disposal of liquid radioactive wastes is usually not carried out.

The radioactive decay of the nuclides in liquid radioactive wastes is accompanied by the release of energy and a rise in the temperature of the geologic medium containing the wastes. There has been some anxiety over the heating of the reservoir to values higher than the vaporization temperature, which would greatly complicate deep injection disposal and create situations ripe for accidents.

Preliminary calculations on the deep injection disposal of liquid radioactive wastes have shown that significant heating can occur with high-level wastes having fission-product activities greater than a few curies per liter. But the hydrostatic pressure means that the vaporization temperature under formation conditions is higher, for example around 220°C at depths on the order of 300 m. For intermediate-level wastes, the temperature rises no higher than some tens of degrees Celsius; for low-level wastes, virtually no temperature rise is seen.

Table 15 shows the energy released by the principal components for a typical waste composition at various times after the formation and disposal of the wastes. Table 16 compares the energy outputs of liquid radioactive wastes destined for disposal and of solidified high-level wastes to be buried in relatively impermeable geologic formations. The energy output declines over time, and the output from solidified high-level wastes is two to three orders of magnitude greater than that from wastes buried in liquid form.

Thermophysical calculations were originally done by a simplified scheme (heat-conduction problem for a bar). Analytical solutions of the conduction equation were derived later, and finite-difference calculations were carried out to model actual conditions as closely as possible.

Table 15. Energy Characteristics of Radioactive Wastes (1 Ci/L)

Nuclide	Half-Life of Parent Nuclide, y	Decay Energy, MeV	Content, %	Energy Output, W/m³		
				0	1 y	10 y
Strontium + ^{90}Y	29.12	1.126	23	0.79	0.77	0.60
Zirconium + ^{95}Nb	0.18	1.655	1	0.04	<0.01	~0
Ruthenium + ^{106}Rh	1.01	1.615	22	1.07	0.54	~0
Cesium + ^{137}Ba	30	0.745	14	0.31	0.30	0.24
^{134}Cs	2.06	1.72	9	0.46	0.33	0.01
Cerium + ^{144}Pr	0.78	1.34	24	0.98	0.4	~0
^{147}Pm	2.62	0.064	7	0.01	<0.01	~0
			100	3.66	2.35	0.85

Table 16. Energy Outputs of Radioactive Wastes

Class of Wastes	Activity, Ci/L	Energy Output, W/m³		
		0	1 y	10 y
Low-level	10^{-9}	$3.66 \cdot 10^{-9}$	$2.35 \cdot 10^{-9}$	$0.875 \cdot 10^{-9}$
Intermediate-level	10^{-5}	$3.66 \cdot 10^{-5}$	$2.35 \cdot 10^{-5}$	$0.857 \cdot 10^{-5}$
High-level	1	3.66	2.35	0.857
High-level, solidified	500	1830.0	1175.0	428.5

Calculations were performed with moving energy sources in the reservoir, hydraulic dispersion, accumulation and retention of nuclides in rocks, kinetic phenomena, radial and vertical conductive heat transport, and other effects. The results showed that formation temperature depends strongly on the energy output of the wastes and the activity per unit volume of the reservoir; this in turn implies a dependence on the nuclide buildup in the rocks, the nuclide composition, and the reservoir thickness. The injection rate chiefly affects the position of maximum heating in the reservoir relative to the well. The thermal properties of the rocks and wastes, the open porosity, the kinetics of the processes, and several other factors are considerably less important determinants of formation heating.

Models described here were studied at various stages of the characterization, design, and operation of deep injection waste disposal sites, the assessment of consequences, and the safety analysis. A comparison of predictions based on the models shows quite satisfactory agreement. These results are examined in Chapter 5. Table 17 gives a general description of the models used to define the processes involved in deep injection radioactive waste disposal.

Table 17. Models of Processes Taking Place in Deep Injection Disposal of Liquid Radioactive Wastes

Process Accompanying Disposal	Consequences of Process	Parameters Governing Process	Mathematical Models, Methods	Scale on Which Process is Manifest
Change in pressure regime of reservoir.	Repressuring dome or cone of depression forms (in case of discharge); direction and velocity of liquid movement change.	Filtration factor, pressure conductivity factor, permeability, regimes and volumes of injection, edge conditions.	Filtration equations; analytic and finite-difference.	Reservoir pressures change over area up to 100 km² during injection.
Filling of reservoir with wastes and subsequent migration of wastes.	Water is displaced from pore space by wastes; wastes mix with subsurface waste; nuclides transfer into rocks by physical and chemical processes; radioactive nuclides decay.	Pressure field, specific capacity (effective thickness and porosity) of formation, dispersion and interphase distribution coefficients, nonuniformity of filtration properties.	Filtration and mass-transfer equations; analytic and finite-difference.	Wastes in reservoir disperse over 1-3 km² during operational period of site, up to 10 km² during maintenance period.
Change in geostatic state.	Formation pressure and stress distribution in geologic medium change.	Pressure field in formation, overburden pressure, physical and of formation.	System of equations for stress state of geologic medium; analytic and finite-difference.	Virtually no effect is seen.
Change in temperature of geologic medium.	Heating zone forms in reservoir.	Waste injection regimes, waste makeup, capacity properties of reservoir, distribution coefficient, heat capacity, thermal conductivity.	Heat conduction equation; analytic and finite-difference.	Heating region forms: 200-300 m² around injection well.

The geologic models and process models used to evaluate the safety of radioactive waste disposal have some special features. For example, the deterministic models described above are transformed to conceptual-probabilistic models where some parameters are replaced by their probability distributions.[57] Such models will be considered in more depth in the next section.

*Protecting
the health
and safety
of plant workers
and the
general public
living nearby
has been
and continues
to be the main
goal of deep
injection
radioactive
waste disposal.*

3.5 Safety Analysis and Environmental Impact Assessment

Protecting the health and safety of plant workers and the general public living nearby has been and continues to be the main goal of deep injection radioactive waste disposal. These concerns have influenced the stages of preliminary development, studies and surveys, design, construction, operation, and postoperational activities. The attainment of this goal is immediately linked to the safety of burial operations. Here "safety" means the ability of the disposal system to prevent or limit the impact of wastes on human beings. Safety is one aspect of a more general property of a system: its reliability. Reliability also includes availability, maintainability, and repairability.[58]

The main safety criteria in radioactive waste handling are governed by the severity of impacts on human beings, both in normal operation of engineering systems and during possible deviations from normality—including accidents—and the probability of occurrence of such accidents. Laws and regulations, including those in force in Russia, together with the recommendations of international agencies, establish dose-based limits on exposure to the population near nuclear facilities (category B).[6] Exposure above the natural background but below set levels is permitted. It is assumed that such exposure will not have marked impacts on human beings in the context of natural and technogenic impacts. Even so, it is recommended that unnecessary exposure be prevented and that exposure be held to the lowest possible level.

The deep injection disposal of liquid radioactive wastes corresponds in the highest degree to the last recommendation, since burial of the wastes would avert any exposure of the population. The population dose burden would thus be zero. The direct impact of deep injection disposal on the public would be restricted use of the subsurface in the burial region; this is not too great an imposition in the vast majority of cases.

Thus the two chief requirements on reliability and safety in deep injection radioactive waste disposal follow from the need to prevent the wastes from having impacts on human beings. First, the dispersal of waste components must be limited to a pre-established region in the geologic medium. And second, uses of the subsurface are restricted within these boundaries. If these tests are met, the wastes will not contact human beings.

These requirements have to do mainly with normal disposal conditions, which include "design" accidents. "Nondesign" accident situations relate to rare natural phenomena such as meteorite impacts or the consequences of human actions (military attack, sabotage, etc.). In such accidents, radioactivity may escape the set boundaries, for example, through the destruction of surface facilities of a working injection well, pipeline, or the like. Even then, however, the affected region will be limited to the sanitary protection area, to which only production personnel are admitted.

Accordingly, the basic safety criterion for the deep injection disposal of liquid radioactive wastes is the localization of the wastes within defined volumes of the geologic medium: the subsurface exclusion zone and the sanitary protection area. In order to verify that this criterion is satisfied at intermediate stages of burial operations, observational data on waste dispersal and other processes are compared with control boundaries and parameters taken from forecasts. Table 18 shows a scheme for generating safety requirements and criteria for the deep injection disposal of liquid radioactive wastes. Compliance with safety requirements based on the stated criteria has been and remains one of the main areas of activity in liquid radioactive waste disposal.

Table 18. Safety Requirements and Criteria for Deep Injection Disposal of Liquid Radioactive Wastes

Principal Objective	To prevent radioactive wastes from having an impact on human beings and their surroundings.	
Basic Requirements	Wastes can disperse only within a predetermined region of the geologic medium; uses of the subsurface within this region are restricted.	
Fundamental Safety Criteria	Scale of waste dispersal within geologic medium must comply with basic requirements; parameters of other processes must satisfy established control limits.	
Basic Steps to Achieve Objective	Investigation, characterization, design	Operation of deep injection disposal system
	Generation of requirements on geologic formations suitable for disposal	Definition of subsurface exclusion zone and sanitary protection area
	Geologic exploration and study to verify compliance with requirements	Monitoring of subsurface waste dispersal and process parameters
	Characterization and design of disposal system so as to comply with basic requirements	Comparison of observations with control limits; inferences about safety of deep injection disposal operations
	Justification of operating regimes; definition of control limits and levels; design of monitoring system	Optimization of operating regimes; monitoring of uses of subsurface in disposal region
	Analysis of hypothetical problems and accidents; development of preventive measures	Implementation of accident responses based on monitoring data

The basic safety criterion for the deep injection disposal of liquid radioactive wastes is the localization of the wastes within defined volumes of the geological medium.

Notions about safety in burial operations were first formulated during the characterization and development of the first injection sites and were similar to the ideas stated here. These ideas were later advanced and supplemented by mathematical modeling with elements of probability theory. The basic methodological issues in safety analysis are examined in what follows.

In performing the safety analysis for a deep injection disposal site, several questions must be answered:

- What is the basis for supposing the wastes will be localized within the prescribed region of the geologic medium?

- What guarantee exists that this requirement will be satisfied, or what is the probability that it will not?

- What will happen if, for any reason, wastes escape beyond the boundaries established; what causes can lead to this effect; can they be remedied?

- Can the burial site be monitored; that is, is it possible to know at any time where the wastes are and what is happening to them?

- Can the wastes in the subsurface be controlled or can their conditions be modified?

- What is to be done after waste disposal operations are concluded; can the disposal site be "mothballed"?

The answers to these questions should be put in terms of the following fundamental factors governing safety:

- the natural isolating properties of geologic formations

- the interactions of wastes with the geologic medium

- the technical decisions taken with regard to configurations, operating regimes, and engineered systems

- the safety of operating regimes together with monitoring and control provisions

- the condition of the geologic medium after the shutdown of the disposal system

The natural isolating properties of the geologic medium depend chiefly on the following:

- the ability of relatively impermeable clay beds in the section to block waste dispersal above the reservoir and the buffer horizon

- slow natural movement of subsurface waters in the reservoir, so that wastes cannot migrate a significant distance from the burial site

- capacity and filtration properties more-or-less uniformly distributed over the reservoir cross section; absence of thin zones of anomalously high permeability that would enable injected wastes to disperse over long distances

- ability of rocks to delay nuclide dispersal.

Predictive calculations based on models of the geology and disposal processes are used in assessing these properties.

In the preliminary phase of the safety analysis, approximate calculations are performed using simplified methods and file data. Qualitative measures of the isolation of geologic formations play a major role in the safety assessment. These include the hydrogeochemical and hydrogeodynamic zonation of stacked beds, the saturation of the reservoir with brines, the absence of active fault zones with planes linking horizons in the potential area of influence, and a low background seismicity.

The provisional safety assessment makes it possible to decide whether to proceed with special geological exploration and studies to obtain needed data for a quantitative safety evaluation. This phase includes simulations and comparisons, which make it possible to establish the scale of waste migration as a function of the geological and operational parameters. The boundaries of the subsurface exclusion zone and the proper restrictions on nearby activities can also be determined.

By statistically analyzing exploration data and identifying the types and characteristics of the probability distributions, it is possible to formulate conceptual-probabilistic models of the geology in which some parameters are expressed by constants and others by probability distributions.[57] Some parameters of the geology display a probabilistic character because of natural variability and unavoidable errors of determination.

The use of such models makes it possible to derive a collection of results, each with a corresponding set of parameters and boundary conditions. Solutions of the predictive problems differ in the probability of realization. The use of conceptual-probabilistic models makes it possible to determine the probability that a given set of geologic conditions will differ from the assumed conditions, and hence the probability of accidents (see below).

The parameters and characteristics to be represented by constants or probability distributions are chosen after an analysis of the variability (uncertainty) and of the degree to which the parameter affects the final results (sensitivity). Sensitivity and variability analyses can also be used to assess the sufficiency and quality of data obtained by geological exploration and studies for use in the safety analysis. If a parameter, say the permeability or hydraulic conductivity, shows a large scatter (variance) about a mean or trend surface, the model of the spatial distribution may not give a good fit to the real picture, or there may be errors in the method of determination.

Experience has shown that the key model parameters influencing the final result are the filtration properties of the reservoir and the relatively impermeable rocks capping it, the thickness and porosity of the latter, the distribution of filtration properties over the reservoir cross section (presence of anomalously permeable zones), and the slope of the natural flow. Of these, the filtration properties usually vary the most; in isolated cases the thickness of the confining formation is the most variable factor. The gradient of the natural flow, on the other hand, can be determined with accuracy and shows little variance. The capacity properties of the reservoir, with allowance for interactions, are characterized by wide variability but have less effect on the safety assessment because the calculations include a wide margin of safety.

Predictive calculations for extreme values of the parameters, including the worst-case combination (whose probability can be determined), make it possible to assess the consequences of disposal operations if the geologic conditions are incorrectly evaluated. The probability of such consequences describes the degree of justification of the forecasts for average values of the geologic parameters and exclusion zone.

A similar analysis of the geologic conditions and exploration results makes it possible to answer other questions: whether claims of waste localization within set boundaries are justified, whether this condition can be guaranteed, and what causes and effects attach to possible errors in the assessment of the geologic conditions.

With regard to interactions of wastes with the geologic medium, the same approaches are used because these processes are described by the parameters of disposal process models, such as effective porosity, distribution coefficient, and degree of desorption of nuclides.

The focus of attention in this interaction analysis is on phenomena that can reduce the level of safety. These include heating of the reservoir by radioactive decay energy, gas evolution due to radiochemical or other processes, concentration of transuranic nuclides in the rocks, and the possibility of spontaneous chain reactions.

Predictions of heating make it possible to set limits on nuclide contents in wastes given the waste composition, reservoir parameters, injection regimes, and accumulation of nuclides in the rocks.

Requirements on waste pretreatment can also be determined, including the maximum allowable activity. These calculations will define the probable ranges of the parameters, so that the quality of the starting data and the reliability of the results can be evaluated. A comparison of calculated and measured data (see Section 5.1) shows satisfactory agreement.

Because radiation-chemical processes under the conditions prevailing in the closed pore space are reversible, with the pressure in the reservoir ranging up to several megapascals, gases formed by radiation-chemical processes are small in volume. This again is confirmed by observational findings. However, gas may evolve rapidly, for example, through the action of denitrifying bacteria. Such processes can be detected in laboratory studies of waste-formation interactions, and they can be forestalled by proper waste treatment. Calculations show that a spontaneous chain reaction would be practically impossible as a result of liquid radioactive waste disposal.

The most complete and coherent safety assessment takes place in the development of the technical-economic justification and design. An environmental impact assessment—including a safety assessment—is done either as part of this phase or as a separate project.

In the course of the technical-economic justification, geologic exploration results and the analysis of oil-field experience make it possible to select the final model and parameters of the hydrogeologic conditions and processes. Predictions can be made for various site conditions. The boundaries of waste dispersal, the pressure fields, the composition, and the temperature can also be determined. This information forms the basis for setting the subsurface exclusion zone and the sanitary protection area and for identifying the environmental protection measures to be taken there.

In this phase of the safety assessment, special value attaches to the analysis of hypothetical problems and accidents, including the listing of such situations, preparation of scenarios and forecasts, evaluation of occurrence probabilities and event trees, and identification of preventive methods. The term "accident" means the occurrence of phenomena or processes, resulting from the deep injection disposal of liquid radioactive wastes, that endanger the health of human beings and the normal performance of their economic activity, harm the ecosphere, and demand urgent responses to arrest them. A "problem" or "complication" is defined as a phenomenon that interferes with ordinary operation of the disposal site and demands special actions.

The analysis of hypothetical accidents usually entails looking at all thinkable and unthinkable scenarios, even implausible ones. This approach is necessary to convince the public, local government, and regulatory authorities that the deep injection disposal of liquid radioactive wastes is safe. (The majority of employees in these agencies are specialists in neither geology nor the protection of subsurface waters.) Hypothetical accidents can be classified by cause as follows:

- accidents stemming from errors in evaluating the capacity and isolating qualities of the geologic medium or from the use of hydrogeologic models fundamentally failing to reflect actual conditions

- accidents that occur when geologic conditions at the disposal site change as a consequence of natural transformations of the geology (including orogenesis, heightened tectonic activity, etc.)

- accidents resulting from the occurrence, in the reservoir or overlying formations, of processes due to disposal operations

- accidents caused by damage to, or deterioration of, wells and surface facilities

- accidents due to human error in the operation of disposal systems

- accidents due to direct action (drilling into the reservoir, driving of mine workings, military action, sabotage)

- accidents resulting from natural disasters such as typhoons, tornadoes, floods, or exceedingly rare events such as airplane crashes or impacts of large meteorites.

Errors in evaluating the geologic conditions can occur when the bedding, porosity, effective thickness, or filtration properties of the reservoir have not been adequately studied. (The filtration properties here include the vertical nonuniformity and features of the natural flow of subsurface waters.) In the reservoir, such errors will result in wastes spreading beyond the predicted limits. Waste migration can occur, for example, via thin beds of anomalously high permeability that were not discovered in preliminary studies. Waste components can also migrate over long distances with the natural flow of the subsurface waters.

If, in the initial stage of disposal operations, phenomena of this type are seen in observational data, appropriate responses include suspension of operations. Accidents involving the impact of radioactivity on the public and biota will not occur if early mitigation measures are taken because the wastes remain localized within the reservoir formation, though over an area larger than was assumed. Geologic exploration conducted to avoid such errors includes prolonged pumping and injection tests with water, in which the reaction of the levels is measured a considerable distance from the central well, as well as tracer studies.

Errors in evaluating the isolating qualities of relatively impermeable horizons overlying the reservoir can lead to wastes penetrating into the shallower buffer horizon. This again is not an accident, because the waste components will remain within the subsurface exclusion zone. The probability of such errors in the evaluation of the filtration properties and isolating qualities is governed by methodology (see later in this section) and lies in the range of 10^{-6} to 10^{-7}, a quite acceptable level.

Faults within the waste dispersal region can occasion vertical movement of the wastes up to shallow horizons if the fault planes are permeable and cut all higher-lying horizons. The vertical migration of waste components along fault zones is, however, quite slow, and the buffer bed serves as spare capacity along the path of vertical movement. As a result, an accident can be averted even though the problems may be quite grave.

Careful geologic exploration incorporating methods described above make it possible to avoid errors in evaluating the ability of a site to isolate wastes and in designing a waste injection facility at the site.

Erroneous design decisions such as locating a disposal site outside the explored area; choosing the wrong intervals, volumes, and pressures; drilling the wrong number of injection and monitoring wells; or establishing the wrong scope of monitoring do not cause accidents. These errors become apparent in the initial phase of disposal site operations. To prevent such errors, technical-economic justifications and designs are examined by independent specialists.

Natural changes in the geology are classified as exogenetic or endogenetic processes. Exogenetic processes wear away the upper parts of the geologic medium and play a significant role in forming the Earth's surface. They are of little significance for reservoir beds lying hundreds of meters or more below the surface. Endogenetic processes are the primary means by which the tectonic and neotectonic structure is formed, causing portions of the surface to experience uplifting and subsidence, and generate earthquakes.

Heightened tectonic activity may reactivate existing faults and create new faults; permeable fault planes can connect horizons and permit the vertical migration of waste components. The increase in activity is not a sudden event but follows a long period of uplifting and subsidence with a significant rise in seismicity. Disposal ranges are located in platform areas where such phenomena apparently will not take place for tens or hundreds of thousands of years to come.

An even longer period of time is required for surface uplifting and erosion to expose a formation containing wastes. Such phenomena will be accompanied by magmatism, the formation of intrusive and extrusive igneous rocks, and the escape of wastes to the surface. In platform areas, these processes fall outside the scope of plausible forecasting. Magmatic activity may even play a beneficial role in localizing wastes in the reservoir, since rising temperature and pressure, magmatic intrusion, and the presence of hydrothermal solutions lead to a marked diminution of sedimentary rock permeability, such as when these rocks are transformed to metamorphic species such as skarns and gneisses. As mentioned earlier, earthquakes have little effect on deep horizons, except when seismic dislocations and seismically active zones appear; liquid radioactive wastes are not buried in such regions.

The deep injection disposal of liquid radioactive wastes involves potentially dangerous processes, a bad combination of which can bring about accidents and complications. If the reservoir temperature rises to the vaporization point under formation conditions (250 to 350°C at depths of 350 to 450 m), a steam-gas phase will appear, and the pressure will rise at the mouths of wells in the central part of the heating area. Because the reservoir is an open system, the wellhead pressure will not exceed the formation or hydrostatic pressure.

An accident comes about if the head of a pressurized well is suddenly damaged, e.g., as a result of sabotage or military attack. Such an accident will contaminate a limited area of the surface near the well, within the first zone of the sanitary protection area. An escape of the steam-gas phase or gas-laden contaminated liquid will be accompanied by a drop in the temperature of the reservoir heating zone; steam formation will decline and ultimately cease. A similar scenario may occur if gas forms in the reservoir, for example, through radiochemical processes.

To prevent the conditions that could cause such accidents, the accumulation and concentration of nuclides in the rocks are controlled by preparing the wastes and pretreating the reservoir formation. The maximum nuclide content in high-level wastes is restricted, and an appropriate injection regime is used for them. Experience in the deep injection disposal of high-level wastes shows that limiting the reservoir temperature to a maximum value is sufficiently effective.

In the safety assessment of deep injection disposal operations, some anxiety is occasioned by the accumulation, in the reservoir rocks, of components that can give rise to explosive conditions. Above all, this applies to transuranic elements, especially plutonium, present in unrecoverable trace amounts in the wastes. Analysis shows that a spontaneous chain reaction requires the formation of a local region or zone in the reservoir containing metallic ^{239}Pu or a highly concentrated solution of this species. Such a zone must be no smaller than 10 cm, and the plutonium medium must not include rocks, formation waters, or waste components. Such conditions are virtually impossible in the reservoir. If a spontaneous chain reaction is assumed, its consequences will be a temperature rise and gas formation, but by no means a nuclear explosion. The standardization of transuranic levels in wastes to be buried affords a further guarantee of nuclear safety.

The dissolution of reservoir rocks by aggressive waste components such as acids may lead to deconsolidation of the filter zone in the formation and shallower-lying rocks adjacent to the borehole. This will result in conditions favorable for leakage via the annulus. Physical and chemical tests in which weakly acidic high-level wastes were injected into formations showed that clay minerals composing 3 to 5% of arenaceous reservoir rocks are susceptible to partial dissolution, feldspars less so. Quartz grains, constituting 70 to 80% of the rocks by volume are not dissolved, only their area may change.

The action of weakly acidic solutions increases the porosity somewhat but does not deconsolidate the rocks in the reservoir. Clay rocks in the relatively impermeable horizons overlying the reservoir are virtually unaffected by weakly acidic solutions because flow processes in these rocks would occur on an insignificant scale. Limestone reservoir rocks obviously suffer greater changes when acid solutions are injected into them. Acidic radioactive wastes are not buried in such formations.

Under unfavorable conditions, the change in formation pressure during waste injection and the associated geodynamic phenomena may cause some complications, such as seismicity in the form of weak earthquakes. These complications do not lead to accidents, damage buildings, or degrade the isolation of the wastes in the reservoir, but they do give rise to complaints by the public and regulatory authorities.

The causes of induced seismicity were examined in Chapter 2. To prevent such effects, the injection site should not be located in seismically active areas or near tectonic faults where the contacting blocks are made of hard rocks under stress. The injection pressure must remain below a maximum value.

Accidents and problems linked with the escape of wastes into shallow beds or onto the surface can take place if the technical condition of the wells deteriorates. Such changes are seen, for example, if the casing becomes leaky and "blowouts" form where a leak in the casing string coincides with an uncemented interval of the annulus. Poor cementing of the borehole-casing annulus, including gaps in the hardened cement or the presence of channels and cracks, permits vertical leakage of wastes and formation fluids through the annulus. The wastes and other liquids can thus enter the buffer beds and sometimes the shallower horizons. The result is contamination of horizons above the reservoir. The scale of this contamination, however, is limited because of the relatively high flow resistance of the leaks. The contamination remains inside the subsurface exclusion zone. Special measures discussed in the next chapter are taken to prevent these phenomena and to detect them promptly should prevention fail.

The failure of surface equipment—wellheads, pipelines, and pumps—can have more serious consequences. This is especially so if equipment is damaged while working and pressurized or if collection and drainage systems are destroyed. The escape of solutions contaminates the surface within the first zone of the sanitary protection area. At the same time, leaks in piping, pumps, and wellhead equipment are design accidents, and design decisions are oriented toward localizing them.

Sudden destruction of pressurized wellhead facilities could allow the release of wastes to the surface, which would continue until the downhole pressure is lowered. The outflow time ranges from hours to weeks and even months, depending on the previous injection regime, the pressure, and the densities of wastes and subsurface waters. A minimal release would result during injection of high-level wastes. The reasons are the brief operational period, the predominant use of free "pouring" injection at a level below the Earth's surface, and the fact that the well is depressurized in the intervals between injection runs. If equipment is destroyed at a well into which intermediate-level wastes are being injected, the release would continue for no more than a few hours, because of the relatively low injection pressure and developed pressure, as well as the high density of the wastes.

Standard provisions at a disposal site include equipment and materials for "killing" a well by filling the borehole with a dense brine to displace the wastes from the borehole into the reservoir. This step may become necessary in the case of injection wells handling low-level wastes, where the volumes and pressures are higher than for intermediate-level wastes, and the density is near or below that of the formation waters (even if they are saline).

Human error in site operations involves incorrect set up of equipment and defects in the assembly of fittings. Human error is considered a cause of accidents. Monitoring systems and sewerage for leaks are ways to prevent such accidents. The most serious consequences can result from injection of unconditioned wastes, especially intermediate- and high-level wastes. If this happens, the reservoir may become overheated, large volumes of gas may evolve, and the problems discussed earlier may occur. To prevent these problems, strict compliance with regulations is required.

Accidents could also result from direct actions taken at radioactive waste disposal sites: for example, drilling and mining into a reservoir containing wastes, or deliberate or inadvertent destruction of working surface equipment. Restrictions on subsurface use in the burial region and enforcement of such restrictions helps prevent unauthorized access to the wastes both during operation of the site and after shutdown. The drilling of deep wells and the driving of mine workings are complicated, time-consuming activities; they cannot be carried out without the knowledge of regulatory agencies. There are no objective reasons for performing such work in disposal sites except in connection with disposal operations.

Sabotage, military attack, and aircraft crashes could cause the destruction of surface facilities. If this happens during waste injection, wastes will contaminate the surface in the first zone of the sanitary protection area. If there is prior warning of military attack or if disposal operations are suspended for a long period, pipes and pumping stations can be drained and wells can be killed with heavy brines. These precautions will prevent the escape of the wastes and the contamination of the surface. After a disposal site has been shut down and the surface equipment dismantled, such actions will not affect the isolation of wastes in the reservoir formation. Once the boreholes have been plugged, the wastes will be contained, they could not rise any higher than the roof of the reservoir.

Natural disasters such as typhoons, tornadoes, floods, and meteorite impacts can damage hard surface installations at a disposal site and cause surface contamination if they occur during the injection period. After the site has been shut down, these phenomena will not affect the wastes.

This safety analysis shows that the possible consequences of hypothetical complications and accidents have very limited impacts on the environment. Effects can be seen only within the subsurface exclusion zone or the first zone of the sanitary protection area of the site, where any activity not connected with disposal operations is banned. Accidents during the disposal phase are very improbable, and their consequences are not comparable in scale with, for example, a nuclear reactor accident involving core damage or the failure of a surface storage pool for high-level liquid wastes.

The consequences of possible accidents and complications during deep injection disposal operations can be assessed on the scale of nuclear power plant events proposed by IAEA in 1990 (JNES). Possible events at nuclear plants are ranked on a seven-point scale.

- Level 7, global emergency of Chernobyl type involving release of a large amount of radioactivity into the environment, with an impact on the public.

- Level 6, grave emergency typified by the 1957 accident at Windscale, England, involving release of radioactivity into the environment.

- Level 5, emergency involving environmental consequences as at Three Mile Island (1979).

- Level 4, in-plant emergency involving exposure of the public; typical case St. Laurent, France (1980).

- Level 3, serious event similar to Vandellós, Spain (1989), in which the public and plant personnel were exposed.

- Level 2, moderately grave event involving a potentially dangerous departure from normal operation.

- Level 1, minor event involving functional excursions of the system that present no danger.

Possible accidents at sites where liquid radioactive wastes are buried will rank no higher than Level 3, with no exposure of the public.

Hypothetical accidents can be assigned to one of two classes by cause and methods of probability assessment:

- Accidents due to subsurface processes developing faster than predicted; processes include nuclide migration, natural changes in the geologic medium, or heating, for example.

- Accidents linked with an unfavorable combination of hard-to-predict random factors; these include failure of a well or surface facilities, human error, and natural disasters.

The probability of the first group of accidents is estimated using conceptual-probabilistic models of the geology. Evaluation of the probabilities in the second group involves experience in the operation of similar engineered systems together with statistical data on natural disasters and catastrophes.

The basic principles used in determining the probabilities of hypothetical problems and accidents, based on the statistical distributions of the model parameters, were discussed earlier. Some supplemental information appears here.

The following expression is used to estimate possible values of the parameters:

$$a_i \pm S_{\bar{a}} \bullet t(f\beta)$$

where a_i = the sample mean of the parameter

$S_{\bar{a}}$ = the sample estimate of the standard deviation of the mean where t = Student's t

β = the confidence probability

f = $n - 1$ = the number of degrees of freedom

n = the number of determinations.

If the parameter has a lognormal distribution, a_i and $S_{\bar{a}}$ are replaced by their logarithms. The interval obtained is then asymmetric about the mean.

The probability of occurrence of an accident is the probability that a model parameter value exists such that the accident will take place. The critical or limiting value of the parameter giving rise to an accident is found by solving an inverse predictive problem subject to the condition that the accident occurs. For example, it is assumed that the components of the waste have reached the human environment in dangerous concentrations or that the reservoir has been heated, and the values of the parameters for which these events do take place are determined. If the probability distribution of a parameter is known, and its main characteristics are the mean $\bar{a}$ and the standard deviation σ_p, the significance level governs the probability of existence of a critical value of the parameter associated with the accident.

The significance level is given by

$$\alpha = 1 - \gamma \int_{-}^{b} (1 - x^2)^{-\frac{f+1}{2}} dx$$

where

$$\gamma = \frac{\Gamma\left(\frac{b+1}{2}\right)}{\sqrt{\pi} \bullet \Gamma\left(\frac{t}{r}\right)}; \Gamma(z) = \int_{0}^{\infty} e^{-y} y^{z-1} dy$$

$$b = \frac{t}{\sqrt{f}}; \quad f = n - 1; \quad t = \frac{a_n - \bar{a}}{S_{\bar{a}}}$$

a_n is the limiting or critical value of the parameter; $\bar{a}$ is the sample mean of the parameter; S_a is the sample standard deviation of the mean of the parameter; and n is the number of determinations of the parameter.

The probability of an accident is taken to be the probability of existence of a limiting value of the parameter on which the development of the accident depends.

Here a_n, $\bar{a}$, $S_{\bar{a}}$ and n are obtained by statistical analysis of parameter values from geologic exploration and physical and chemical studies, or by analysis of conceptual-probabilistic models of the geology and the processes involved in disposal, including information about the probabilistic characteristics of the parameters. If the probability distributions of the parameters are approximated by a lognormal or similar distribution, a_n, $\bar{a}$ and $S_{\bar{a}}$ are replaced by their logarithms. It is possible to identify not just one but several parameters that govern the development of accidents. If the parameters are independent, the limiting value is found for each in turn, and the probability of its existence is found with all the other parameters held constant. The maximum probability of an accident due to the possible simultaneous deviation of the parameters from the mean values used in the predictions is given by the expression

$$P = 1 - \prod_{i}^{m} (1 - \alpha_i)$$

where α_i is the probability of existence of limiting values of the I-th parameter.

The probability values are then compared with analogous figures for alternative technologies for handling wastes from the chief plants. This procedure makes it possible to draw conclusions about the relative danger or safety of disposal operations.

At this stage of technical-economic justification and design, predictive calculations are also done for anomalous conditions. In this way, the causes of accidents can be refined, the possible consequences can be assessed, needed measures can be taken to prevent them and remedy their consequences, and the effectiveness and cost of these measures can be evaluated. The same data are used for creating the monitoring system; here special attention is devoted to identifying the initial features and preconditions for the development of complications (problems) and accidents.

Safety analyses relating to the deep injection disposal of liquid radioactive wastes from particular facilities are presented in Chapter 5.

The main task in assessing and ensuring safety at the construction and preoperational phase is to monitor the quality of all work, particularly the construction, casing, and cementing of wells.

Features of the geologic structure are refined as operational wells are drilled; documentation is developed using the results of geologic and geophysical studies and tests. In many cases, the parameters of the geologic model and subsurface processes can be corrected, and supplementary predictive calculations can be performed. The results are used to prepare the operational rules for the disposal site and to justify the scope and nature of monitoring. Safety assessment during the operational phase of a disposal site is based on the results of monitoring observations carried out in accordance with the rules.

Without discussing observational methods in detail here (see Chapter 4), note that the principal monitored parameters are the distribution of waste components in the reservoir and the variations in formation pressure and temperature that accompany waste dispersal. Tracking the formation pressure makes it possible to detect trends in waste dispersal long before they become apparent in the observation well, so that disposal operations can be optimized.

The results of monitoring observations are compared with predictions made in the design phase and during preparation of the operational rules. Predictions are generated for various times (e.g., 1, 3, 5, 10, 15, 20 years) after operation begins. If the measurements agree with the forecast values, the disposal process is considered normal. If actual values differ from predicted values, causes for the discrepancy are sought, and a decision is made on how to optimize disposal operations.

Control actions include varying the injection regime (turning the injection and relief wells on and off). This makes possible some degree of control over the direction of waste movement in the reservoir.

Changing the waste composition during preparation for disposal makes it possible to control the transfer of nuclides to the solid phase or the desorption of nuclides by the moving formation fluid. Setting up the proper temperature schedule in the reservoir can increase the accumulation of nuclides in the rocks and enhance the bonding strength. For example, raising the temperature disrupts soluble compounds in the system comprising wastes, rocks, and formation waters and creates poorly soluble compounds that trap and coprecipitate the nuclides. The temperature, in turn, is regulated by the activity of the wastes, the acidity or alkalinity of the medium, and the volume and rate of injection.

Antimigration curtains can be used to restrict the nuclide dispersal. These curtains are formed in the reservoir by treatment with special reagents and gas mixtures.

In the postshutdown phase of a disposal site, safety is ensured by the following means:

- Enforce restrictions on subsurface uses within the exclusion zone and waste localization region, for a specified period of time.

- Dismantle engineered systems (wells, pipelines, surface installations) and seal wells.

- Rehabilitate the surface within the first zone of the sanitary protection area.

- Monitor the subsurface in the context of federal or regional environmental surveillance.

Additional study of the geologic medium and the disposal processes during the operational period of disposal ranges makes it possible to obtain a considerable amount of information that can be used to improve forecasts of waste behavior underground. Given these data, a safety analysis for the postshutdown phase is carried out during shutdown planning; at the same time, the boundaries of the subsurface exclusion zone are refined (the zone is reduced in size), as are the subsurface monitoring systems. Monitoring disposal sites in the postshutdown period will make it possible to ensure the requisite safety in the use of natural resources.

4 Deep Injection Disposal Systems for Liquid Radioactive Wastes

Deep injection disposal systems, also called burial ranges or deep injection storage sites, include surface and subsurface equipment, monitoring and control subsystems, and portions of the subsurface within the exclusion zone and the sanitary protection area. The main functions of a disposal site are to receive pretreated wastes; inject wastes into wells; monitor waste volume and composition, injection conditions, waste dispersal into the reservoir, and the condition of the geologic medium; and optimize operations to secure the maximum degree of safety and produce the minimum environmental impact. Figure 9 is a schematic diagram of an injection disposal site.

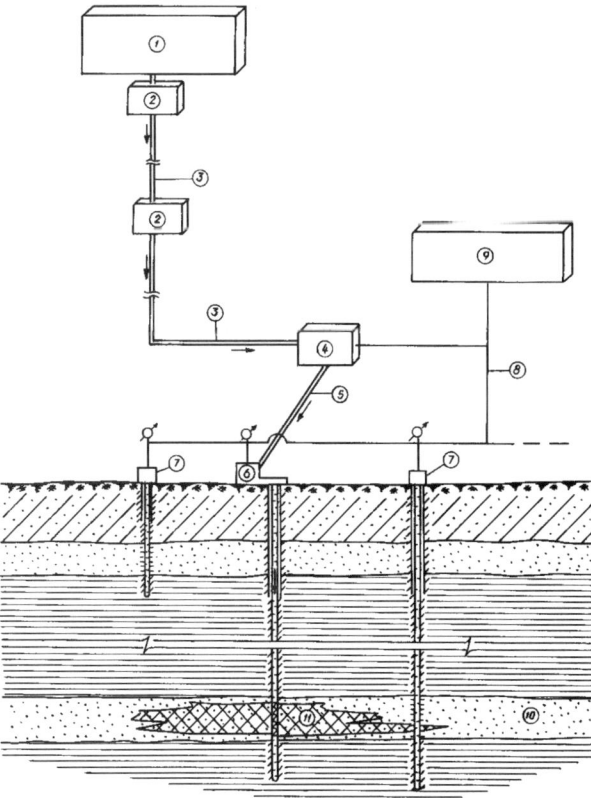

Figure 9. *Schematic of a Deep Injection Disposal Site for Liquid Radioactive Wastes. 1 - treatment plant, 2 - transfer pumping station, 3 - low-pressure pipelines, 4 - holding tank building and injection pumping station, 5 - high-pressure pipeline, 6 - injection well, 7 - observation monitoring well, 8 - collection of monitoring data, 9 - administrative and engineering building, 10 - reservoir formation, 11 - wastes in reservoir.*

The heart
of the
engineered
system at
any deep
injection
disposal site,
whether
for liquid
radioactive
wastes or
for industrial
wastes, is
the system
of wells.

The management of a disposal site has much in common with that of an oil field, but the great potential hazard presented by wastes, along with requisite special procedures, means that the methodologies and design solutions adopted for waste disposal are quite different. The distinctions relate chiefly to the justification of the systems, safety practices, technical requirements, the design of the basic structures (wells drilled for various purposes), monitoring practices, and the standards for justifying and implementing deep injection disposal.

The first injection and monitoring wells, constructed in the early 1960s, were similar to oil wells in design and surface (wellhead) equipment. These wells were located at the experimental liquid radioactive waste disposal site at Area 18a of the Siberian Chemical Combine (SCC). Later facilities had significantly different equipment and structures, which were specially designed. Development of these designs followed the preparation, coordination, and approval of the technical and economic justification. Detailed development of engineered systems took place at the working design stage. The layout of injection and monitoring wells, the disposal conditions, and the monitoring intervals were based on calculations and simulation (Chapter 3) with initial data from geologic exploration and laboratory tests.

This chapter briefly describes the most important elements of disposal systems and practices. This information is based on experience in creating and operating existing disposal sites, along with other developmental work. The main aspects covered include engineered systems, monitoring methods, and the current and former standards that form the legal basis for establishing and operating disposal sites.

4.1 Engineered Systems for Disposal Sites

The heart of the engineered system at any deep injection disposal site, whether for liquid radioactive wastes or for industrial wastes, is the system of wells. The efficiency and safety of burial depend largely on the design and operating conditions of the various types of wells. Operational wells in disposal systems differ in several ways from exploration, oil, and water wells, particularly in the expected reliability, the importance of keeping horizons isolated from one another, construction materials, and the monitoring of well status.

Surface equipment at a disposal site consists of the wellhead equipment, piping, pumping stations, electric power and heating systems, instrumentation and automation systems, office and engineering buildings, and roads. The surface facility also includes mobile devices and equipment, the laboratory, and means of transportation.

Operational wells are subdivided into three main groups: injection, instrumented (monitoring), and relief. In addition, disposal sites may construct special wells, for example to obtain process water from shallow horizons or to inject grout when repairing or shutting down operational wells.

There are two distinct types of disposal systems, with and without relief of the reservoir formation. Pressure relief, or the pumping of pure water from special wells, is used to lower the formation pressure when waste injection tends to raise the pressure. Relief is employed, for example, at the liquid radioactive waste disposal site of the Mining and Chemical Combine (MCC), where the reservoir is limited in plan so that high formation pressures are generated by the injection of wastes. When a site is operated without relief, no pure water is removed by pumping.

Injection patterns can be classified as circular (example: Area 18 at the SCC), linear (sites at the MCC), and "cluster", which is intermediate between these [as at Area 18a of the SCC or the pilot waste disposal site of the Scientific Research Institute of Nuclear Reactors (NIIAR)].

Injection wells and some monitoring wells at deep injection disposal sites for liquid radioactive wastes have one distinguishing feature: a multistring or "telescopic" construction in which all the casing strings are brought to the wellhead, and the uppermost horizons of shallow-lying fresh waters are provided with a minimum of two casing strings. The inner string, which will contact the wastes, must be made of corrosion-resistant materials. The borehole-casing and intercasing annuli are cemented from the casing shoe to the wellhead; this practice must ensure reliable separation of horizons from one another. Workmanship in well construction is carefully checked, as are the conditions of all structural elements.

Features that may be present in injection wells include the conductor and/or surface casing strings, the intermediate and production strings, the screen, and the tubing string. Figure 10 shows various designs of wells. The conductor, sometimes called surface casing, reinforces the upper horizons of loose rock during the drilling process. The conductor extends to a depth of 20 to 25 m and functions only while drilling is in progress. This string is not used in stable rocks. The surface casing plays a similar role but is usually run to a greater depth, sometimes as much as 100 to 150 m. The surface casing can likewise be omitted in stable rocks. The conductor and surface casing are commonly between 200 and 600 mm (8 and 22 in.) in diameter.

The intermediate string strengthens and isolates the uppermost horizons of the section and ensures successful penetration of the reservoir as well as installation and proper cementing of the production string. Local geology and technical conditions largely

Workmanship in well construction is carefully checked, as are the conditions of all structural elements.

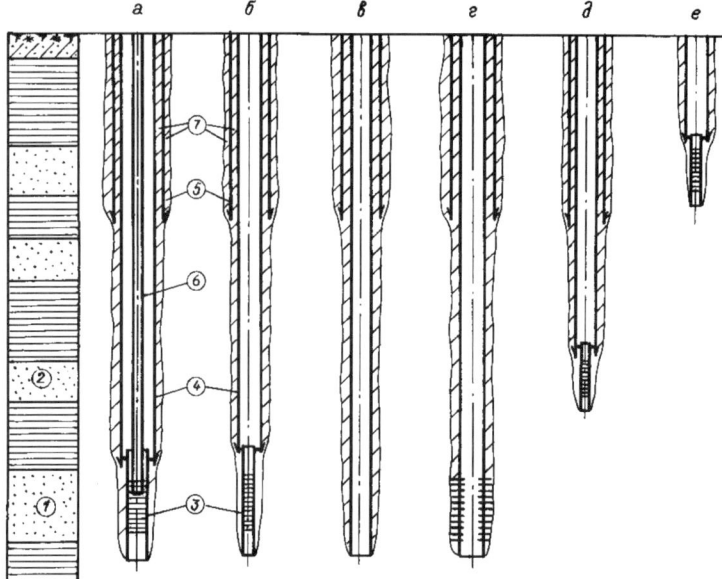

Figure 10. *Construction of Wells at Disposal Sites for Liquid Radioactive Wastes. (a) injection well, observation well, (b) well with filter zone, (c) geophysical (blind) well, (d) exploration well converted to monitoring service, (e) well reaching buffer horizon, (f) well reaching shallower horizon, 1 - reservoir formation, 2 - buffer horizon, 3 - screen unit, 4 - production string, 5 - intermediate string, 6 - tubing, 7 - cement.*

govern the depth of the intermediate string. These conditions include the presence of unstable intervals in the section or lost-circulation zones (zones where drilling mud or cement is absorbed). Shutting off these intervals with the intermediate string makes it possible to easily pass through the hole below the production string and to install and reinforce the production string.

The main string in the well is the production string, the interior of which contacts the wastes. This string is run to the lower part of the confining bed above the reservoir formation (which is fitted with a screen). If the filter zone of the well, by which the interior of the well communicates with the reservoir, is made by perforating the appropriate interval of the production string, that string is run beyond the floor of the reservoir. The intermediate and production strings are usually between 150 and 300 mm (6 and 12 in.) in diameter.

The screen (filter string) serves to reinforce the reservoir in the exposed interval, feed wastes into the formation, and prevent finely divided formation material from entering the well. The string can be perforated in the reservoir interval (this is ordinarily done at the surface), or a separate filter element can be used. This filter or screen can be wire-wound, ceramic, or a porous synthetic material.

The lower part of the well, in the interval of the horizon beneath the reservoir, is a sump formed by the continuation of the screen or production string. This region collects finely divided solids either passing from the formation into the well or already present in the wastes.

Tubing strings are used to bring wastes to the screen zone and diminish the contact of wastes with the internal surface of the production string. Tubing is commonly used when wastes are highly aggressive in relation to the casing pipe material. The bottom portion of the tubing can be fitted with a packer to isolate the wastes from the annulus between the tubing and production strings, which is then filled with an inert liquid.

For trials, an injection well can be fitted with an instrumented string along the well axis inside or alongside the tubing. The interior of this string does not communicate with the interior of the injection well. The instrumented string is filled with pure water permitting geophysical observations directly in the injection well.

The condition of an injection well and its compliance with applicable standards are determined to a great extent by the drilling method, which in turn is governed by the local geology and technical conditions. An essential aspect of well drilling is monitoring the progress, curvature, and cross section of the hole; the installation and cementing of the strings; the tightness of the pipe and the welded or screw joints; the condition of the cement; and the nature of the hole penetrating the reservoir. To these ends, well construction is accompanied by tests and observations that include the following: monitoring dynamic progress characteristics, the mud and cement parameters and velocities, and the pressure and flow rate of cement; performing geophysical tests both in the open hole (electric logging, caliper logging, inclinometer survey) and after the well is shut-in (thermometry, radioactive logging, acoustic and nuclear cement bond logging, flowmeter logging). The requirements on well condition, drilling and casing methods, logging, completion, testing, and acceptance of wells are discussed in the design documentation.

Observation (monitoring) wells, the second type, are used to observe waste dispersal, processes accompanying dispersal, and the condition of the geologic medium. Wells that penetrate the reservoir, and in which waste components may appear, are usually called "observation" wells, while wells used for tracking the condition of overlying or underlying horizons where wastes are not anticipated are usually referred to as "monitoring" wells. This division is rather arbitrary, and in practice both kinds of wells are likely to be called by either name.

The design and construction of monitoring wells are similar to those of injection wells, though monitoring wells generally have fewer strings and thus smaller diameters. In wells that are close to the injection contour and that penetrate the reservoir, the intervals of shallow groundwater horizons are commonly provided with two casings. Distant wells and monitoring holes have the surface string or intermediate string run to a shallower depth. Standby injection wells, which serve as monitoring wells in the first operational period of the site but are later used as injection wells, are most similar in design to injection wells.

Monitoring wells include a blind or geophysical type whose interior does not communicate with the reservoir. Such wells permit geophysical measurements without direct contact between the downhole instruments and the wastes. Wells of this type are usually close to the injection contour of the disposal site, where levels of radioactive nuclides may be high. Monitoring wells drilled into shallow freshwater horizons are often called "sanitary-hydrogeologic" wells.

As in the case of injection wells, the design, construction methods, measurements, and condition requirements for monitoring wells are governed by specifications. The roster of monitoring wells generally includes previously drilled exploration wells provided they comply with requirements.

The third type of well used for deep injection disposal is the relief well. The purpose of relief wells is to relieve the pressure in the reservoir formation when necessary or, equivalently, to pump pure water out of the reservoir when the formation pressure must be lowered for waste injection. Since wastes will not appear in relief wells, less stringent requirements apply to their design. Downhole electrical pumps are used to transport water, though an airlift may be used. The designs of relief and injection wells are governed by the geological and technical conditions relating to construction and operation. In essence, they are no different from the well designs used in other fields.

In their journey to the reservoir formation, wastes are first received at the surface facility where they will be prepared for disposal. Pretreatment is most commonly carried out at the production plants or in treatment buildings. The wastes are then allowed to settle in holding tanks, usually at the disposal site.

Wastes are transported from the treatment plant to the injection station via low-pressure pipelines and transfer pumping stations. The design and hardware used in pipelines and pumping stations depend on the activity and composition of the wastes, the injection mode, and the daily flow volume. Process wastes are carried in double-wall corrosion-resistant piping laid in trays lined with corrosion-resistant materials and flexible plastic. Leak-catching vessels with level alarms and drains are installed along the pipeline route. The requirements on this equipment are set forth in departmental standards for the design of radiochemical plants.

The injection or pumping station is one of the key surface installations. It is located within the disposal site near the waste injection contour.

The following types of pumps are used for waste injection:

- For low-level nonprocess wastes, centrifugal pumps with delivery heads ranging from 20 to 25 atm gauge up to 50 to 60 atm gauge are used. These pumps have stuffing boxes. Multistage types include the MS, AYaP, MD, and other series; PN, KVN and other feed pump models are also used.

- For process wastes, BEN and TsNG series of glandless pumps deliver injection pressures ranging from 10 to 12 up to 25 atm gauge.

Injection stations feature two- or three-zone designs meeting the standards of radiochemical plants.

The pipes carrying wastes to the wells are designed in the same way as the pipelines transporting wastes to the disposal site. Design features include piping laid in trays, leak collection, and double-wall pipes.

Wellhead equipment comprises freestanding half-buried installations (pavilions) housing the wellhead connections. The plan dimensions of injection well pavilions are 4 x 4 m, 6 x 6 m, and 6 x 9 m. Automatic heating systems and balanced ventilation with air cleaning are installed in these pavilions. Fittings here include valving and regulators as well as instrumentation with local and remote readouts. Two distinct types of injection wellhead installations are used for valving and regulation. The first type is located directly on the wellhead, the second 5 to 10 m away. The second type of installation makes it possible to set up, for example, a well-repair derrick directly over the wellhead without dismantling the pavilion. The design of the wellhead systems also depends on the activity level of the wastes.

The wellhead connections of an injection well must direct the wastes into the hole, regulate and monitor the pressure and flow rate, allow the connection of supplementary pumping or cementing equipment, and catch leaks from the fittings and convey them to a holding vessel for later injection. The fittings must be rated for a pressure at least twice the specified injection pressure.

The wellhead connections of monitoring wells are much simpler than those of injection wells. Fairly high concentrations of waste components will appear in the downhole liquid in monitoring wells located close to the injection contour; the wellheads here are also under pressure. Small-diameter sampling pipes can be passed through the wellhead fittings, run to the screen depth and sealed

The first task of disposal monitoring is to ascertain from two sources the real distribution of waste components in reservoirs, and in the geologic medium as a whole.

at the top. In this way, samples can be taken and the pressure measured while allowing as little contamination as possible.

The surface equipment becomes yet simpler for monitoring wells located farther from the injection wells, for blind (geophysical) wells, and for wells penetrating shallower horizons.

Standard water-well equipment is usually suitable for the structure and surface fittings of relief wells. The pure water produced by relieving the reservoir formation is discharged into sewerage or used for processing. If this water is mineralized, it can be disposed of in "lost circulation" horizons above or below the reservoir that also contain mineralized waters.

The surface complex of a deep injection disposal site also includes electrical and heating systems, office and engineering buildings, laboratory apparatus, mobile equipment, transport, and communications.

4.2 Monitoring of Deep Injection Disposal

In accordance with the foundational principles, the deep injection disposal of liquid radioactive wastes must be accompanied by monitoring. The composition and volume of the wastes, their dispersal in the subsurface, and other associated processes must be overseen. Monitoring data must be used to check earlier predictions and to demonstrate safety.

The first deep injection disposal site for liquid radioactive wastes was the experimental range at the Siberian Chemical Combine. Special monitoring wells were provided, and observations and measurements were performed. All the techniques used were borrowed from geologic exploration practice and the petroleum industry. The methods included hydrodynamic observations, sampling of formation fluids from wells for various kinds of analysis, and geophysical logging (gamma logging, thermometry). Monitoring systems currently in use at disposal sites include several subsurface monitoring functions. Development of hardware continues.

The monitoring of liquid radioactive waste disposal has several unique features. The large subsurface volume that may be affected by disposal operations requires many monitoring points. Because the wastes are liquid, their position changes almost continuously, and vertical redistribution within the reservoir becomes probable. As a consequence, sizable amounts of monitoring data must be collected. The potential hazard presented by the wastes makes it necessary to regularly acquire information on the position and condition of the wastes. Such data must be made available to regulatory authorities and the public.

The first task of disposal monitoring is to ascertain from two sources the real distribution of waste components in reservoirs, and in the geologic medium as a whole. The first source is direct determinations of the components in the formation liquid and in the reservoir rocks; the second is based on indirect indicators: the variation in formation pressure (potentiometric surface of subsurface waters), the temperature, and the resistivity of water in the well. These results are compared with predictions, and inferences are made about the nature of the processes. If necessary, measures are taken to optimize operating conditions—or, in the extreme case, to prevent accidents.

A no less important task is to track the condition of the wells, which largely governs the effectiveness of disposal site operations and the reliable isolation of the waste-containing reservoir from shallower horizons and the surface. Well condition monitoring is done both by determining operating parameters and by direct examination while running downhole instruments into the well. The results are used to plan and schedule preventive maintenance and to define the expected life of the well. The condition of surface equipment is also monitored.

Two essential aspects of disposal monitoring are injection parameters (pressure and flow rate) and waste composition. Of special concern are the levels of certain components in the wastes; these determine the effectiveness and safety of disposal operations. Injection parameters and waste characteristics must comply with stated requirements.

A special function of the monitoring system is the early detection of developing problems. Early detection and appropriate responses make it possible to forestall a serious problem or an accident.

The possible occurrence of extremely improbable events is also monitored. Two such events are seismic vibrations associated with disposal operations and a change in the position of the surface. Table 19 describes the concerns, tasks, and methods of deep injection disposal monitoring.

The basis of the monitoring system is a network of special wells with instrumentation and equipment, means of transport, and specialist staffs. The number and disposition of these wells are set in the design phase. Criteria for this determination include the need to acquire representative data at regular intervals, to track the position of wastes in the subsurface, and to detect early the conditions that would lead to problems and accidents. The network of wells is denser at the injection contour so that maximum data can be obtained about processes in the early stages of operation, and corrective action can be taken when necessary. The wells are also spaced closer together where the geology is complex, for example, at benchlike and grabenlike structures in

A special function of the monitoring system is the early detection of developing problems. Early detection and appropriate responses make it possible to forestall a serious problem or an accident.

Table 19. Monitoring of Deep Injection Disposal of Liquid Radioactive Wastes

Monitoring Concerns	Monitoring Tasks	Monitoring Methods	Parameters and Characteristics Determined	Criteria for Interpreting Monitoring Data
Geologic medium (subsurface)	Ascertain distribution of radioactive wastes and progress of disposal processes.	Radiochemical testing	Content of radioactive waste components in subsurface waters	Compliance of radioactive waste distribution with predictions; absence of wastes in overlying horizons and beyond boundaries of subsurface exclusion zone.
		Hydrodynamic method	Potentiometric height of surface of subsurface waters	
		Geophysical method (radioactive, electric logging; thermometry; seismic observations)	Characteristics of geophysical fields in wells and at surface	
Condition of wells	Determine operating parameters and condition of structural elements; verify isolation of reservoir from higher-lying horizons.	Radiochemical testing	Content of radioactive wastes in subsurface waters of shallower horizons in region of well	Characteristics of shallower horizons do not differ from background; levels of subsurface waters in higher-lying horizons do not differ from natural values.
		Hydrodynamic method	Potentiometric height of surface of subsurface waters in shallower horizons in region of well	
		Geophysical method (acoustic, radioactive logging; thermometry).	Characteristics of geophysical fields in intervals higher than reservoir	Parameters do not differ from background; acoustic or density characteristics of cement comply with requirements.
Injection conditions and waste composition	Verify compliance with stated requirements, check condition of filter zone of well.	Direct measurements, chemical and radiochemical analyses of waste samples	Flow rate and pressure of injection; contents of limiting components in radioactive waste	Design values and regulatory standards.

the basement or where the content of sands in confining beds increases.

Blind wells (for geophysical measurements) and monitoring wells drilled to shallow horizons are also located in the immediate vicinity of the injection wells. The total number of monitoring wells is quite large: 259 in the two injection areas of the Siberian Chemical Combine, 70 in the two sites of the Mining and Chemical Combine, and 32 in the pilot test area at the Scientific Research Institute of Nuclear Reactors. Surface monitoring stations and networks are also located inside and outside the disposal site. These stations carry out standard observations of air, water, soil, and vegetation.

The instruments and apparatus used for monitoring are the same as used at radiochemical plants, in geologic exploration, and in oil fields: pressure gauges, flowmeters, samplers, level gauges, geophysical instruments, logging instruments, and laboratories. Many special devices have also been developed specifically for deep injection disposal sites, including:

- the "Kedr" well-logging system[a] used to measure beta and gamma radiation downhole, even at high temperatures and high gamma exposures

- a 12-channel seismic monitor.

Disposal site monitoring comes under site operations (a service or division of a service), which is under the waste processing division of the producing plant. The environmental protection service also carries out observations in off-site wells and some on-site wells.

Monitoring of the deep injection disposal of liquid radioactive wastes can be subdivided by concern:

- monitoring waste injection: this includes determining the volumes and compositions of the wastes, the total quantities of substances and radioactive nuclides disposed of, the injection parameters, and the operating regimes of the injection wells

- monitoring subsurface conditions: including observations of the distribution of waste components in the formation and the progress of associated processes, among them the pressure and temperature regimes of the reservoir and shallower horizons, the characteristics of emitted radiation, the composition of formation fluids, and seismic conditions

The total number of monitoring wells: 259 in the Siberian Chemical Combine, 70 in the Mining and Chemical Combine, and 32 in the Scientific Research Institute of Nuclear Reactors.

(a) "Kedr" (Cedar) is the brand name. The abbreviation SNRK in the original probably stands for "continuous radioactive logging system."

- monitoring the condition of engineered systems, chiefly the wells and surface facilities.

Monitoring can also be categorized by location. Process monitoring is implemented right at the disposal site and aims to determine the location of the wastes, the condition of the wells, and the operating regimes. Geologic monitoring deals with the surrounding geologic medium.

The scope and character of process monitoring will depend in turn on the well's location within the disposal site. The greatest volume of such observations is carried out near the contour of the wastes in the reservoir, and in regions where the rocks are heated most, for example. After the front of the wastes has passed a monitoring well, the volume of monitoring there decreases.

Geologic monitoring consists of downhole observations outside the boundaries of the disposal site and in wells penetrating higher-lying groundwater horizons inside the site. There are other objects of attention around radiochemical plants: active or potential sources of subsurface contamination (radioactive waste lagoons, solid waste burial sites, pipelines, etc.) and consumers of subsurface waters. To address these concerns, a "unified system of subsurface monitoring" is often established, tasked to detect contamination of subsurface waters, define the areas of influence of various sources, identify measures needed to stop contamination, and optimize water resource management. The system also includes monitoring wells at the disposal site in addition to the process monitoring wells. Organizationally, the unified subsurface monitoring system comes under subdivisions of the plant responsible for monitoring the environment, which increases the objectivity of the acquired data.

The scope and methods of monitoring depend on the operational phase of the disposal site. Observations are numerous in the early stages, since disposal and site models must be refined and operating regimes must be corrected, while the regulatory authorities must be convinced that disposal is safe. Later, the volume of monitoring can be decreased. In the final operational phase of the disposal site, the volume of monitoring increases again so that starting data can be acquired for use in planning the postshutdown phase.

Monitoring includes the following observations and determinations:

- Hydrogeochemical and radiochemical studies involve sampling the subsurface waters and the filtrate of wastes from the wells, analyzing their compositions, and determining the levels of waste components. Sampling is performed with special apparatus that makes it possible to obtain representative data.

- Hydrodynamic observations involve determining the potentiometric level of subsurface waters in the reservoir formation and higher-lying horizons (measurement of wellhead levels or pressures). Because hydraulic perturbations propagate much faster than injected solutions in water-bearing horizons, hydrodynamic observations make it possible to establish the overall trend in waste dispersal long before the wastes appear in a monitoring well; the isolation of the reservoir from overlying formations can also be verified. Standard instrumentation is used to determine levels and pressures.

- Geophysical observations include the downhole measurement of physical fields set up by waste disposal. Gamma logging, in which a radiation sensor is moved along the borehole axis, yields the distribution of gamma-emitting radionuclides in the rocks just beyond the borehole wall. The value of this information lies in the fact that it reveals the vertical hydraulic nonuniformity of the reservoir and the condition of the confining beds. Beta logging also makes it possible to measure the activity of the borehole liquid over the section, without having to bring samples to the surface. Thermometry or thermal logging enables workers to track the variation in temperature due to heat released by radioactive wastes and detect any vertical redistribution of these species. Resistivity logging allows inferences about the salt content of the downhole liquid, while flowmeter logging gives a profile of liquid "absorption" over the section of the reservoir. Standard and special instrumentation is used for geophysical investigations. The interval between measurements is based on predictive calculations and experience. Figure 11 illustrates a monitoring scheme with examples of the results obtained.

The geophysical measurements are used in constructing graphics that reflect subsurface conditions: maps showing equipotential lines, maps with isopleths of activity and radioactive waste component levels in the formation liquid and the rocks, isotherm charts, and geologic and hydrogeologic sections plotted with monitoring data. These materials form the basis for refining the model parameters by solving inverse problems, and also for correcting and optimizing the disposal regimes.

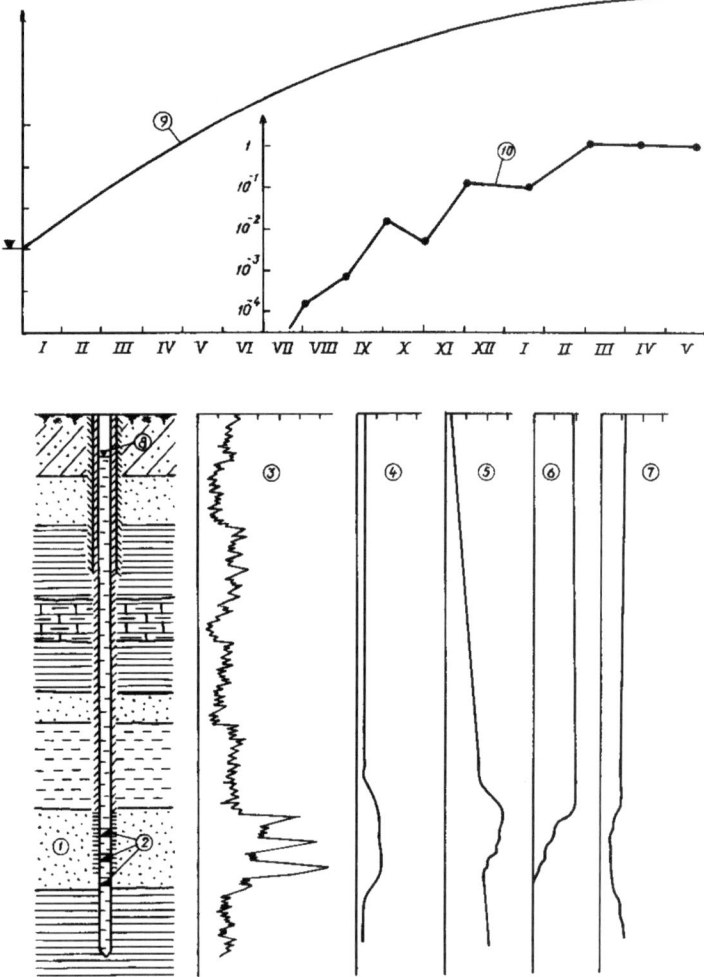

Figure 11. *Monitoring Data for a Disposal Site. 1 - reservoir formation, 2 - liquid sampling intervals, 3 - gamma log, 4 - beta log, 5 - temperature log, 6 - flowmeter log, 7 - resistivity log, 8 - level of subsurface waters in well, 9 - time variation of subsurface water level, 10 - time variation of levels of waste components in well.*

4.3 Legislative and Regulatory Framework for the Deep Injection Disposal of Liquid Radioactive Wastes

Liquid radioactive wastes pose a great potential hazard, and their deep injection disposal is a critical activity entailing large expenditures and making heavy use of natural resources and subsurface regions. This kind of activity is subject to regulation by the state, based on special documents and checks to be sure their requirements are met.

Experience dating from 1955 shows that standards and regulations in this field can be arbitrarily classed in three basic groups.

The first includes legislative acts and similar documents having the force of law. These contain fundamental requirements for the protection of human beings and the environment, along with principles governing deep injection disposal. Such documents are permanent in nature.

The second group comprises standards, rules, principles, and specifications implementing the legislated requirements. They give concrete values of the various parameters and characteristics, and methods and technologies for achieving these values. These documents can be corrected and supplemented as necessary.

Documents of the third group are process standards and codes along with methodologies developed for each plant or enterprise and each type of operation or study.

The three groups of regulatory and legislative documents differ in the procedure and level of preparation, approval, confirmation, and enactment. If the creation and adoption of legislative acts are the prerogative of the highest organs of government, standards are developed and adopted mainly by departments, sometimes with the advice and even the leadership of interested bodies such as the Academy of Sciences and the National Committee on Radiological Safety. Standards relating to production are created by enterprises or by institutes and are usually approved by subdepartments.

The development and approval of the legislative and regulatory framework for deep injection radioactive waste disposal took a certain amount of time and a corresponding scientific effort. In the late 1950s and early 1960s, when a quick solution was needed to the deep injection disposal problem, government decrees and orders took the place of laws. These documents were based on preliminary scientific results evaluated by the scientific and technical councils in the ministries of medium machinery construction, public health, and geology, with Academy of Sciences experts taking part. The first directive regarding deep injection disposal of liquid radioactive wastes came out on September 13, 1958 (No. 3019rs) and was signed by Prime Minister A. N. Kosygin. This decree covered the main tasks and requirements, set forth basic organizational principles, and allocated responsibilities by department.

Because environmental protection was so important, even in deep injection disposal systems for liquid radioactive wastes, government decree No. 1036 was adopted in 1959—*On Strengthening State Supervision of the Use of Subsurface Waters and on Steps to Protect Them.* At the same time, a document called *Principles on the Procedure for Using and Protecting Subsurface Waters Within the USSR* (1960) was developed. All work on creating deep injection disposal systems was carried out under these principles.

Sections 19-21 of the *Principles* state, "...It is forbidden to discharge wastewaters containing radioactive substances into injection wells or subsurface pits. In exceptional cases, after special investigations have been performed, this may be permitted by the Ministry of Public Health of the USSR jointly with republic-level geology and mineral resource management agencies after the question has been agreed on between the latter and the USSR Ministry of Geology and Mineral Resource Management."

Work on the radioactive waste injection problem was entirely in accord with the above requirements as far as organization and content were concerned. For example, geologic exploration aimed at the justification and selection of a geologic structure and reservoir formation for the reliable and safe disposal of liquid radioactive wastes was carried out by a specialized organization of the Ministry of Geology, called PGO (Production Geological Association) "Hydrospetzgeologiya." Official reports by this agency formed the basis for developing the design documentation for disposal sites. Specialists from the territorial geological administrations of the Main Geological Administration of the Russian Soviet Federated Socialist Republic (RSFSR) took part in on-site work. The method of deep injection disposal for liquid radioactive wastes was approved in principle by the College of the USSR Ministry of Public Health in 1960. Specialists from the Third Main Medical Administration and the Institute of Biophysics of the USSR Ministry of Public Health took an active role in developing public health aspects.

Accumulated experience in geologic exploration and research on the waste injection problem, together with the scientific generalization of data on the geologic structure, tectonics, hydrogeology, and minerals, provided specialists of PGO Hydrospetzgeologiya with a basis for a 1965-70 project. Under the leadership of Academician Sidorenko, these experts prepared and issued the *Predictive Map of Hydrogeologic Conditions for the Deep Injection Disposal of Industrial Wastewaters in Deep Aquifer Complexes*. This map covered the USSR at a scale of 1:2,500,000. The Ministry of Geology had the map examined by the College of the USSR State Committee on Science and Technology. In its decree No. 99 of April 1, 1970, the Committee approved the work that had been done and advised Gosplan (the State Planning Committee) and ministries and departments to consider the possibility of deep injection disposal for wastewaters that did not lend themselves to treatments. This option was to be considered in planning the construction of new enterprises and the reconstruction of existing ones.

The Scientific Council of Engineering Geology and Soil Science and the lithosphere section of the Academy of Sciences Scientific Engineering Council on Biosphere Problems noted in their own decrees that the Predictive Map contained a scientific assessment of the entire USSR from the standpoint of the possibility of injecting industrial wastes into the subsurface.

Activity in the deep injection disposal of liquid radioactive wastes was reflected in the *Fundamental Sanitary Rules for Working with Radioactive Substances and Radiation Sources*. The OSP-72/87 code now in force includes conditions for the deep injection disposal of liquid radioactive wastes in Section 9.8.

Later, several legislative acts and standards were developed and adopted dealing with deep injection of industrial wastes and wastewaters. These include:

- *Fundamentals of Water Legislation*, 1970 (Articles 14, 15, 31)

- *Water Code of the RSFSR*, 1977 (Articles 74, 75, 76)

- *Fundamentals of USSR Legislation on Mineral Resources*, 1975 (Articles 9, 12, 14, 27 and others)

- *RSFSR Code on Mineral Resources*, 1976 (Articles 37, 38, 44, 64, 66 and others)

- *Instructions on the Procedure for Approval and Issuance of Decisions on Special Water Utilization*, 1978 (Section 3.7.8)

- *Principles of Groundwater Protection*, 1984 (Articles 47-51)

- *Instructions on the Procedure for Allocating the Subsurface for Purposes Not Related to the Extraction of Minerals* (1984)

- *Provisional Technical Orders for the Design of Surface Facilities and Utilities at Sites for the Deep Disposal of Liquid Wastes* (1979).

The year 1979 saw the development of *Provisional Sanitary Rules and Specifications for the Construction and Operation of Sites for the Deep Injection Disposal of Liquid Radioactive Wastes*, designated VSP and TU P3-79. The issuing agencies were the All-Union Scientific Research and Design Institute of Industrial Technology (VNIPIpromtechnologii), the Institute of Biophysics and the Third Main Medical Administration of the Ministry of Public Health. Currently there is a permanent regulatory document: *Sanitary Rules and Specifications*, SP and TU EKH-93, which is based on experience in operating and investigating injection sites.[5]

For nearly 14 years, VSP and TU P3-79 has been the basic and sole standard in this country governing the creation and operation of deep injection disposal sites for potentially hazardous and highly toxic liquid industrial wastes. The document was approved by the Ministry of Geology, the State Inspectorate for Mining Safety, the USSR State Committee on Hydrometeorology and Environmental Monitoring, the Ministry of Public Health, and the State Committee on Atomic Energy. It has been used to design many deep injection disposal sites in other industries, including those at the "Orgsteklo" and "Korund" manufacturing plants at Dzerzhinsk (Nizhegorodskaya oblast), the PKhZ Chemical Plant at Pervomaiskii (Khar'kov oblast), the Kirovo-Chepetsk Chemical Combine (Kirov oblast), and the Cheptsa Machinery Factory (Udmurt Republic). Each site is designed to include basic provisions for the safe and secure disposal and localization of wastes for which there are no industrially tested methods of purification and reprocessing. The possibility, safety, and economic expediency of building the sites have been demonstrated through comprehensive studies and technical and economic justifications.

The new *Environmental Protection Act* passed by the Supreme Soviet of the Russian Federation, and published on March 3, 1992, contains a number of new requirements and restrictions on the discharge of wastes and sewage (Article 54.3), the deep injection disposal of potentially hazardous and highly toxic wastes (Articles 54.4, 54.6, 45.3), and the conduct of state ecological assessments (Articles 36.1 and 36.2).

Article 54, Section 3, forbids the discharge of wastes and wastewaters into aquifers used for other purposes ("common use"). According to the *Commentary on the Environmental Protection Act*, published in 1992 by the printing office of the Supreme Soviet, this language does not exclude the use of reservoir formations for the deep injection disposal of toxic wastes.[59]

Three documents of the Russian Federation—the *Mineral Resources Act* of February 21, 1992 (Article 23.8), the *Procedural Rules for Licensing of Uses of Mineral Resources* of June 15, 1992 (Sections 13 and 14), and the *Instructions on Application of the Procedural Rules for Licensing of Uses of Mineral Resources*—reflect in a clear and definite way the questions of deep injection disposal of wastewaters, deep injection disposal of toxic and highly hazardous industrial wastes, the conditions for performing these operations, the procedures for licensing, and the implementation of the deep injection disposal technology.

The Siberian Chemical Combine, Mining and Chemical Combine, and Scientific Research Institute of Nuclear Reactors have received the licenses for using geological formations for deep injection of liquid radioactive waste.

The use of subsurface regions for the burial of wastes involves a fee. The amount of the fee is set by the *Procedural Rules and Conditions for the Collection of Fees for the Right to Use Subsurface Zones, Aquifers and Seafloor Regions* (Section 19), affirmed by decree of the Russian government on October 28, 1992 (No. 828).

Currently, a number of bills and standards are in preparation that would also deal with the deep injection disposal of radioactive wastes. Thus, the creation and operation of deep injection disposal systems for liquid radioactive wastes are covered by a suitable legislative and regulatory framework that has lost none of its value in the light of newly issued documents.

A hotly debated topic in connection with the deep injection disposal of liquid radioactive wastes is how long the isolation of the wastes in the subsurface can be guaranteed. The requirement of "permanent" isolation is virtually unrealizable, since in the long run all geologic formations are subject to transformations. It is extremely hard to forecast how society will develop over millennia or to what extent geologic formations containing radioactive wastes may become involved in the sphere of human activity. For these reasons, there is a limit on the maximum period over which the condition of radioactive wastes in geologic formations can be predicted. The Nuclear Waste Policy Act of 1982, for example, states that forecasts of the groundwater flow and mass transport parameters are needed for 1000 and 10,000 years into the future.[60]

5 Geologic Conditions and Results of Operating Deep Injection Disposal Systems for Liquid Radioactive Wastes

Each of Russia's existing sites for the deep injection disposal of liquid radioactive wastes is unique because of the quantity of radioactive substances contained in the wastes, the problems to be solved to protect human beings from radiation, and the scale of impacts on the geologic medium. The practical experience gained from setting up and running these liquid radioactive waste disposal facilities, along with experiments and observations, can be used to refine and correct theoretical constructs and models developed to handle environmental protection problems at these sites, then predict the contamination of groundwaters and allow the burial of toxic wastes at other candidate sites. Even though the theory is far advanced, some aspects share the same short-coming: little or no validation has been carried out.

This chapter deals with three deep injection disposal sites for liquid radioactive wastes. They are located at the Siberian Chemical Complex (SCC) at Tomsk-7, the Mining and Chemical Complex (MCC) at Krasnoyarsk-26, and the Scientific Research Institute of Nuclear Reactors (NIIAR) at Dimitrovgrad. Information presented here is based on the most recent publications.[61, 62] The material is organized in a manner that will 1) elucidate the conditions and features governing the safety of deep injection disposal, and 2) scrutinize the results of many years' operations at these sites. Site geology is discussed first since this is a major safety factor and also determines how specific problems in exploration, research, and design will be solved. Surface and downhole systems, injection regimes, and monitoring techniques are discussed as well. Results are presented for comparison with predictions made at various stages of the operations. The final topic will be those difficulties that arose either when the sites were established or when wastes were buried and that still exist today.

5.1 Disposal Sites at the Siberian Chemical Complex

5.1.1 Geology

The geology of the SCC disposal site was studied in special exploration projects and was refined when the operational wells were drilled and investigated. The site is located in the articulation zone of the West Siberian Basin and the Sayan-Altai folded region, where hard rocks of the Paleozoic basement lie beneath sand-clay strata of Mesozoic-Cenozoic rocks containing sandy beds with reservoir properties and relatively impermeable clay confining beds.

The Paleozoic basement rocks crop out at the surface 25 to 30 km south of the disposal site on the outskirts of Tomsk. West, north-west, and north, they are buried at depths of 350 to 450 m below the site (Figure 12). These Paleozoic basement formations are chiefly composed of slightly metamorphosed shales with a clay weathering crust in the upper portion; the sand-clay strata lie above this. Figure 13 is a stratigraphic column of the Meso-Cenozoic sediments, indicating the age indexes and arbitrary nomenclature of the horizons.

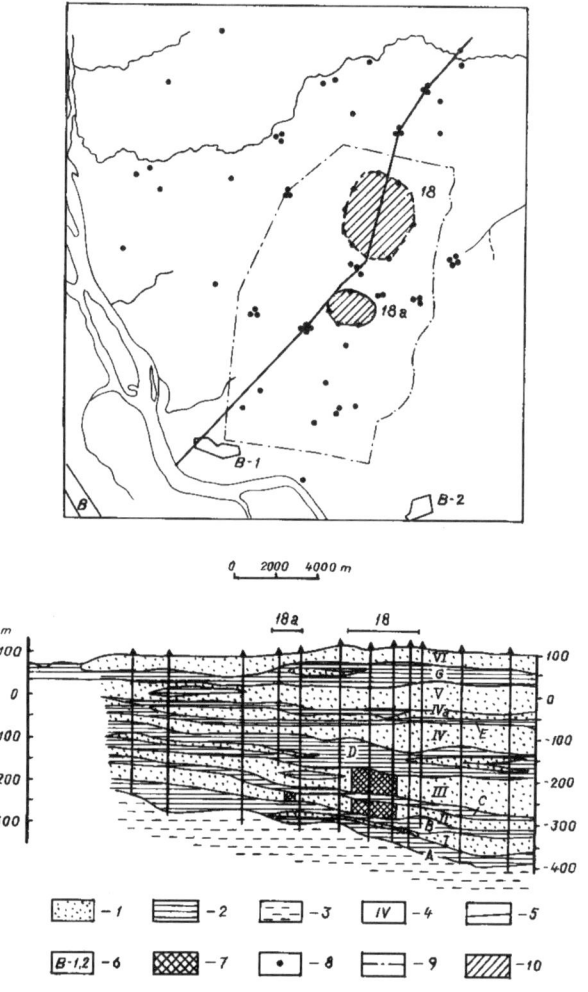

Figure 12. *Map of Disposal Sites (Areas 18 and 18a) at Siberian Chemical Combine, Geologic Section of the Region. 1 - permeable rocks, 2 - relatively impermeable rocks, 3 - relatively impermeable rocks of paleozoic basement, 4 - arbitrary index of horizon, 5 - line of section, 6 - wells producing subsurface waters from shallow horizons, 7 - portion of formation occupied by wastes, 8 - monitoring wells, plant, 9 - boundary of subsurface exclusion zone, 10 - areas of disposal sites.*

ERATHEM	SYSTEM	SERIES	STAGE	SUITE	SYMBOL	ARBITRARY SYMBOL	LITHOLOGY	THICKNESS	DESCRIPTION OF ROCKS
CENOZOIC	QUATERNARY	Holocene / Pleistocene			Q	VI		50-60	Sands with lenses of pebbles and gravel with intercalated loams, sandy loams and clays
	NEOGENE	PLIOCENE	Upper	Kochkorskaya	N_1kc			10-55	Clays with lenses of sand, sand with gravel and pebbles
	PALEOGENE	OLIGOCENE	Upper	Azharminskaya	P_3agr	G		1-32	Clays with intercalated sands
			Middle	Novomikhailovskaya	P_3nm			0,5-80	Clays with lenses and intercalated beds of sand, lignite
			Middle	Atlymskaya	P_3at	V		35-74	Sands with pebbles and gravel, with intercalated clay, lignite
			Lower	Yurkovskaya	$P_{2-3}jur$	F		2-18	Kaolinized plastic clays, locally sandy-silty
		EOCENE	Middle	Lyulinvorskaya	P_2ll	IVa		0,5-26	Clayed sands
			Lower	Kuskovskaya	P_2kk	E		8-10	Finely divided clays
MESOZOIC	CRETACEOUS	UPPER — SENONIAN	Danian Maestrichtian Campanian Santonian Coniacian	Symskaya	K_2ams	IV		40-80	Sands with intercalated clays
				Symskaya	K_2ams	D		20-47	Clays with intercalated sands
			Turonian	Simonovskaya	K_2amn	III		37-94	Sands with intercalated clays
			Cenomanian			C		10-20	Compact clays
						II		25-40	Sands with intercalated clays
		LOWER	Albian	Kiiskaya	K_1ks	B		20-35	Kaolinized clays
			Aptian			I		18-48	Sands with intercalated clays
			Barremian Hauterivian	Kiyalinskaya	K_1ks	A_2		Up to 74	Specked clays with intercalated conglomerates, sandstones
			Triassic Jurassic		T-J	A_1		Up to 64	Weathering crust, white kaolinized clays
Paleozoic	Carboniferous	LOWER	Visean	Basandaiskaya Lagernosidskaya Yarskaya	C_1bs C_1lg C_1jr			160	Silty-clayey shales, slates or shists with intercalated sandstones with diabase and lamprophyre dikes

Figure 13. Stratigraphy of Paleozoic and Meso-Cenozoic Sediments in the Vicinity of the SCC

A grouping within the sand-clay strata, based on all geologic attributes, distinguishes a complex of Cretaceous sediments (represented by sand horizons I, II, III, and IV and clay horizons A, B, C, and D) and a complex of Paleogene and Quaternary sediments (comprising sand horizons IVa, V, and IV and clay horizons E, F, and G). The boundary between the Cretaceous and Paleogene sediments corresponds to horizon E, which separates horizons IV and IVa.

The reservoirs used at the site are sand horizons II and III, which are underlain by relatively impermeable clay horizons B and A and the clays of the Paleozoic weathering crust. The sand horizons II and III are covered by the relatively impermeable D horizon, which occurs throughout the region. The deeper-lying sand horizon I, which separates horizons B and A, is not present everywhere. Horizons II and III are separated by clay horizon C, which is arenated and features an increase in permeability in the southeast portion of Area 18. Above horizon D is horizon IV, which serves as a buffer bed relative to the host formations II and III and is overlain by relatively impermeable horizon E.

Reservoir horizons II and III are made of medium-grained sands varying in the degree of clayiness. The most typical minerals are quartz (70 to 80%); feldspars of orthoclase, microcline, or plagioclase type; mica- and hydromica-group minerals; and clay minerals of the kaolinite and montmorillonite groups. Carbonate minerals and organic matter are also seen.

The relatively impermeable horizons are represented by various clay rocks: speckled, compact, and "fat" clays, sandy-silty clays in some places, and sideritized clays. Fissuring is seen locally, although it is atypical.

The hydrogeologic stratification differs from the age-based stratification already described; this point is important for the safety analysis. Two aquifer complexes are identified in the section: the lower one, including Cretaceous horizons I, II, and III, and the upper one comprising Cretaceous horizon IV and Paleogene and Quaternary horizons IVa, V, and VI. These complexes are separated by horizon D of relatively impermeable clay sediments, which has confining qualities. The complexes are distinguished on the basis of several features: the difference in pressure head, hydrogeochemical characteristics, the geologic-geophysical correlation, and the helium concentration field, for example. The isolating properties of horizon D are established both by analyzing the natural pressure fields and through extended pumping and injection tests and monitoring of site operations over a span of years.

A substantial difference is seen between the flow properties in the upper and lower parts of the section. While the transmissivity of horizons IV, V, and VI ranges up to 1600 m^2/d, the value for horizons I, II, and III is 60 to 80 m^2/d. The transmissivity of horizon IV lies between 80 to 200 m^2/d. The high permeability of the upper horizons accounts for their widespread use for water supply outside the subsurface exclusion zone and the sanitary protection zone of the disposal sites. The limited amount of water present in the lower aquifer complex makes it unpromising for these purposes.

The horizons of the lower aquifer complex contain pressurized waters of bicarbonate-calcium composition; the mineralization is 0.3 g/L. Potentiometric levels reach 300 to 320 m above the roofs for horizon II and 250 to 280 m for horizon III. The hydrogeologic parameters of the reservoir horizons, determined by pumping and injection tests, are presented in Table 20. (Geologic exploration and borehole tests show that the Paleozoic formations have low permeability and do not contain reservoirs; for this reason, they were not studied in detail.)

Table 20. Flow and Capacity Parameters of Reservoir
Formations and Horizon D in Areas 18 and 18a

Parameter	Unit	Area 18a Horizon II	Area 18 Horizon II	Area 18 Horizon III
Depth	m	314-341	349-386	270-320
Thickness	m	30-50	30-50	50-90
Effective thickness	m	13-30	13-24	22-75
Total porosity	—	0.35	0.35	0.4
Effective porosity	—	0.05-0.14	0.1	0.15
Transmissivity	m^2/d	17-24	24	34
Hydraulic conductivity	m/d	0.7-0.9	0.5-3.0	0.2-2.2
Pressure conductivity factor	m^2/d	$1 \cdot 10^5$	$1.2 \cdot 10^5$	$2 \cdot 10^5$
Pressure head over top	m	300-320	325-350	250-280
Thickness of horizon D	m	28-29	58-62	58-62
Hydraulic conductivity of horizon D	m/d	$1.2 \cdot 10^{-4}$	$1.2 \cdot 10^{-4}$	$1.2 \cdot 10^{-4}$

The SCC is located in the southeast part of the Ob' River artesian basin. The discharge area for waters of the lower aquifer complex is confined to the confluence of the Ob' and Tom' rivers. Waters in the lower aquifer complex naturally move south to southwest with velocities of 3 to 5 m/y. The natural slope of the potentiometric surface of the subsurface waters is 0.0009.

Important disposal safety factors are the position and characteristics of the relatively impermeable horizons that overlie and underlie the reservoirs that will be used for disposal. At the SCC, the use of the lower aquifer complex for disposal and of the upper aquifers for water supply makes this question somewhat pressing, even though the water wells are located outside the sanitary protection zone and the subsurface exclusion zone.

Taken as a whole, the relatively impermeable clay horizons identified in the geologic section of the SCC area feature inconsistent composition, the presence of sand interbeds, and a decreasing thickness (wedging out) to the southeast, which is due to the reduction of the section in connection with a decreasing depth of the basement. Regions of possible interconnection between horizons are, however, found a significant distance (10 to 12 km) from the sites and outside the area affected by waste disposal.

An arenated portion of horizon C at the southeastern boundary of the Area 18 disposal site, acting as a flow window, was discovered by geologic exploration. This window has virtually no effect on the waste localization conditions, since at Area 18 wastes are emplaced in reservoir horizons II and III, which are divided by horizon C. Drilling data and filtration tests do not reveal any such zones in horizon D within the disposal sites, nor do hydrodynamic observations during disposal operations. The hydraulic conductivity of horizon D was 1.20×10^{-4} m/d according to extended monitoring.

The area where horizon E wedges out to the southeast is also fairly distant from both disposal sites and water wells. (Horizon E separates the upper aquifer complex, used for water supply, from the lower complex.) Contamination of water wells via this region is virtually impossible even though there may be a change in levels in horizon IV due to rapid withdrawal of waters from horizons IVa and V when the water wells are operated.

During geologic exploration, attention was focused on the nonuniformity of flow properties in the reservoir horizons. Geophysical investigations, including tracer studies, did not reveal zones or intervals with anomalously high flow properties; their absence was later confirmed by monitoring while operation was under way. It was, however, assumed that solutions could flow at a high rate through a thin interval if the wastes were injected at a high pressure (higher than the hydrofracturing pressure of the formation). For this reason, the maximum waste injection pressure was limited.

According to available data, the Paleozoic basement in the SCC region shows only a slight degree of tectonic fracturing. The largest fault is thought to constitute the echelon-like zones along the left bank of the Tom' River, outside the SCC site boundary. A flexure-like feature is seen in the sedimentary rocks in the northeast part of Area 18; the continuity of beds is unbroken, corresponding to a bench in the basement, evidently tectonic in origin. Special flow tests did not establish any interconnection between horizons in the area of the flexure adjacent to the disposal site. A lineament was identified in Area 18, corresponding to a bench in the surface relief of the Quaternary sediments; this feature does not extend to a great depth.

Abyssal fault hypotheses based on an aeromagnetic survey of areas adjoining the SCC were not confirmed by test boreholes on the SCC site. Drilling did reveal dikes of igneous rocks in the Paleozoic formations; these are usually the main factors giving rise to magnetic anomalies.

Recent and Quaternary movements of the crust in the SCC region involve, on the whole, uplifting at a velocity of 0.5 to 1.2 mm/y. These movements will not substantially alter the hydrogeology of the region in the next few thousand years. Natural seismicity in the region has a rating of less than 6.

These findings make it clear that the geologic structure of the SCC region permits the deep injection disposal of liquid radioactive wastes. The main arguments in favor of this conclusion follow:

- The section contains bedded porous sandy rocks (reservoir formations) whose capacity and flow properties are sufficient to allow injection of the specified volumes of liquids at pressures below the hydrofracturing pressure of the formation.

- The reservoir rocks are underlain and overlain by relatively impermeable clay rocks that can act as confining beds. These confining beds are present nearly everywhere in the potential area affected by waste disposal.

- There is convincing evidence that horizon D separates the upper aquifer complex, which is influenced by water wells, from the lower aquifer complex, which is to be used for waste disposal.

- The groundwaters typically move at velocities of 3 to 5 m/y, and such values guarantee localization of the wastes for extended times within the territory of the SCC and in the sanitary protection zone and subsurface exclusion zone.

- The flow properties of the reservoir horizons, as a whole, are uniform in section and in plan, so that there is no reason to expect much nonuniformity in filling the host formations with wastes or in the formation of "tongues" or intervals of anomalously high flow velocity that would greatly affect the areal distribution of wastes.

- No tectonic structures have been found in the vicinity of the SCC that could bring about the vertical interconnection of horizons and the anomalous migration of wastes.

- The low-salt levels in the formation waters, along with the presence of clay minerals in the formation rocks, mean that the migration of waste nuclides will be delayed by physical and chemical processes.

- The geologic structure of the SCC region and the hydrogeology were studied in the course of special geologic exploration work that included drilling several wells and conducting geophysical investigations and flow tests (among other studies, see Section 3.2). The resulting geologic data were quite representative and permitted responsible decision making.

The first rough forecasts of waste distribution in host formations, arrived at in the final stage of the geologic exploration, indicated

The geological structure of the SCC region permits the deep injection disposal of liquid radioactive wastes.

that the area occupied by nonprocess wastes (50 million m³) in horizons II and III would be 7 km², corresponding to a circle 1.5 km in radius. The area occupied by 5 million m³ of process wastes in horizon II would be 2.6 km², corresponding to a circle 1 km in radius. Possible later displacement of wastes through natural movement of subsurface waters was estimated to be on the order of 5 km per thousand years. These areas and distances lie within the site boundary of the producing plants. The results of the geologic exploration gave no grounds to suppose that any anomalous phenomena could arise from the geologic structure and lead to rapid dispersal and migration of the wastes.

5.1.2 Waste Disposal Sites

Areas 18 and 18a, used for the deep injection disposal of liquid radioactive wastes, are located near the main plants and within the sanitary protection zone of the SCC. Other waste-handling facilities are also located in the sanitary protection zone: a building with cleanup equipment, exposed storage pools, industrial ponds, and solid waste repositories.

Because of the serious hazard potential of wastes and the vital nature of related decisions, a justification and design was first performed for the experimental site on Area 18a, which includes five injection wells in the center of the site, four monitoring wells in an inner monitoring array (125 m from the site center), and five in an outer monitoring array (380 to 400 m from site center). Exploration wells located 1200 to 1400 m from the center were also used to determine the subsurface conditions. Figure 14 shows the layout of injection and monitoring wells in the central portion of the site.

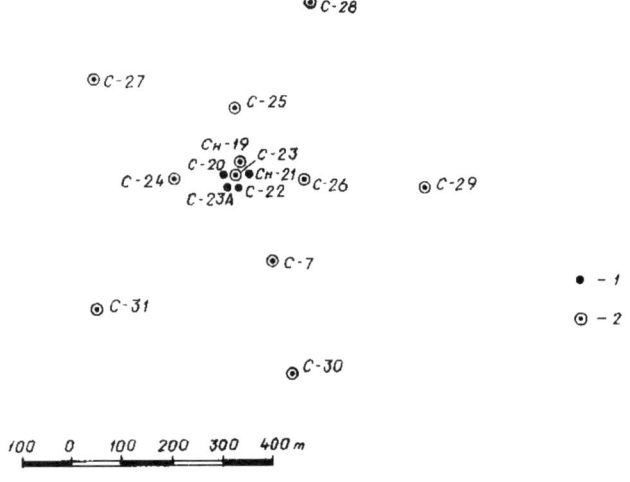

Figure 14. *Locations of Injection Wells and Monitoring Wells in the Experimental Disposal Site. 1 - injection well, 2 - monitoring well.*

The initial plan called for the emplacement of 110,000 m³ of intermediate-level process wastes from the open storage pools. The main objective of the experiment was to verify, under field conditions, the compatibility of the wastes with the geologic medium, the stability of injection well operation, the predicted changes in formation pressures, the filling of the host formation with wastes, and the performance of the surface equipment.

For most of the run, the injection pressure was 1.1 to 2.2 MPa, and the flow rate was between 300 and 1000 m³/d. Changes in the pressure head due to waste injection were recorded in the instrumented wells. By way of example, Figure 15 shows the pressure heads (repressuring dome) for the wells of the experimental disposal site.

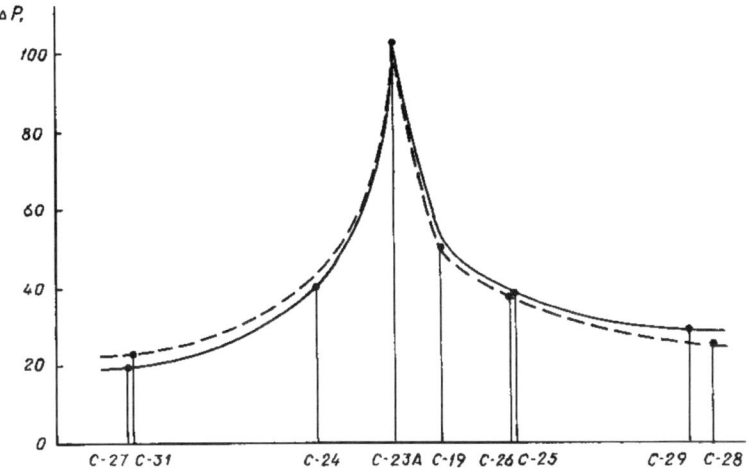

Figure 15. Level of Subsurface Waters in Horizon II Relative to Static Level When Radioactive Wastes Were Injected Into Well 23a of the Experimental Site

The decantate from the storage pools primarily contained salts of sodium (nitrates, acetates, carbonates, sulfates), and silicic acid. The total salt level was 160 to 200 g/L, with a specific activity of 10^{-2} to 10^{-1} Ci/L. The radionuclides present included strontium, cesium, ruthenium, and cerium. The formation liquid was sampled in the monitoring wells and analyzed, and in this way the regular appearance of waste components in the wells was established. The time variation of the specific activity of the waste filtrate in the wells is shown in Figure 16.

The rise in activity during the period from June to August 1963 was due to the passage of wastes and subsurface waters through the well area. The periodic decline in the activity of the formation liquid is associated with pauses in injection. The activity in the formation liquid is one to two orders of magnitude below the activity in the wastes. The reason for this difference is that some nuclides are held up through sorption in spite of the high-salt level in the wastes.

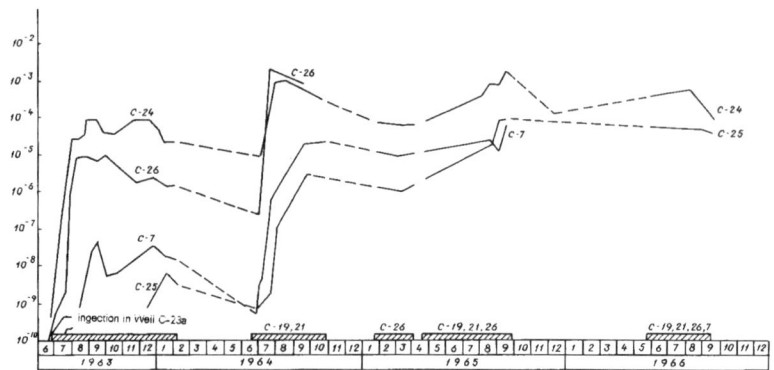

Figure 16. *Specific Activity of Formation Liquid in Monitoring Wells of the Experimental Disposal Site*

The results, taken as a whole, confirmed ideas about waste dispersal and made it possible to refine some initial values that were later used in the design process. Positive results achieved in the deep injection disposal of process wastes at the experimental site led to an extension of its operation to 1975-1980. Some of the wells were shut down as planned, while others remained open for monitoring service. The total volume of wastes emplaced at the experimental site was 2.1 million m³.

The experimental results made it possible to plan and construct an experimental site for the disposal of low-level nonprocess wastes (Area 18) and a full-scale site for process wastes (Area 18a). In view of the large volumes of nonprocess wastes emplaced at Area 18, up to 6300 m³/d, the site had to be redesigned. Two rows of type "N" injection wells (11 wells) were drilled at the site center; these penetrated horizons II and III. Monitoring wells and standby injection wells were arranged concentrically around the injection contour (some 100 wells). As shown in Figure 17, 26 discharge wells are located on the exterior boundary of Area 18. The distance from the site center to the boundary (design waste dispersal contour) is between 1.6 and 2.0 km. The site has an area of about 10 km² and an effective host-formation volume (horizons II and III) of 60 million m³.

Selection of the site layout was preceded by experimentation and modeling of the operating regimes on a mesh-type differential analyzer. This work was performed by specialists at the All-Union Scientific Oil and Gas Research Institute. The discharge (relief) contour was designed to correct the repressuring dome and lower the waste injection pressure and also to control the movement of the waste contour. The results from the early years of site operation, however, showed that the formation pressure rose only slightly in the central part of the site and at the heads of the injection wells. This is so because injection was periodically interrupted for technical reasons, pressure redistribution occurred, radionuclides were effectively held up by the rocks, and the formation proved to have a greater specific capacity than predicted. These factors meant that pressure relief for the host formation was a less urgent problem than predicted.

The Area 18 pilot disposal site was placed in service in 1967. The maximum injection pressure was set at 2.5 MPa for horizon II and 2.0 MPa for horizon III. The total design injection rate was taken as 6300 m³/d; the actual figure did not exceed 5500 m³/d.

For the first 8 to 10 years, wastes were injected into each horizon through two to three wells simultaneously. The rate was nonuniform, and there were periodic pauses. The injection rate varied between 1000 and 2500 m³/d for horizon II and between 1000 and 3000 m³/d for horizon III. As the filter zones of the host formations became plugged by suspended solids in the wastes (colmatage), the injection rates declined and the number of wells operated at the same time was increased. The dimensionless resistances of the filter zones, characterizing the quality of each well with respect

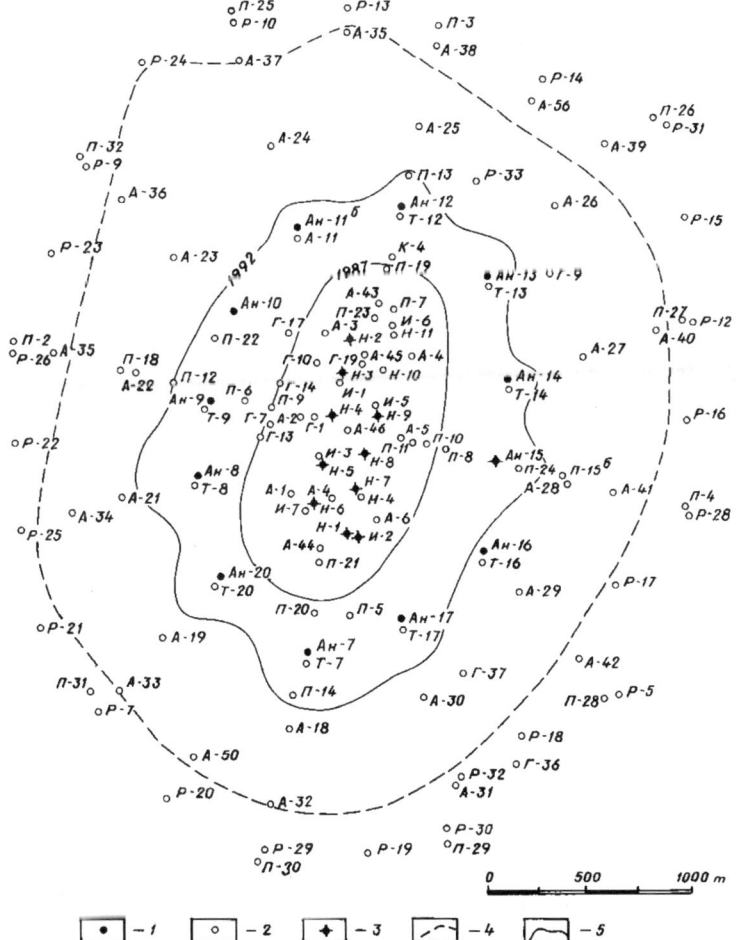

Figure 17. *Disposition of Wells and Waste Disposal Contours at Area 18. 1 - injection well, 2 - monitoring well, 3 - shutdown wells, 4 - design waste dispersal contour, 5 - contour of low-level wastes.*

to the way the formation was exposed, increased from initial values in the "units" range to 90 in the final stages of well operation. The total volume of wastes injected into each type N well was somewhat over 1 million m³; wells N-1, N-7, and N-4 received over 3 million m³. When their capacities had been exhausted and their technical conditions were deteriorating, the type N wells were shut down as planned, and standby type AN injection wells were put in service. Only these wells have been used for injection since 1989.

Waste injection created a repressuring dome in Area 18. The formation pressure in the central part of the area is 10 to 15% higher than the natural value. Figure 18 shows the distribution of the repressuring dome in the horizon II reservoir of the Area 18 site as measured in monitoring wells; a theoretical curve is plotted for comparison.

The radioactive wastes disposed of at Area 18 contain nitrate and sulfate ions as well as detergents. The salt content does not exceed 20 g/L, the pH ranges from 6 to 10.5, and the total specific activity ranges from $n \times 10^{-8}$ to $n \times 10^{-6}$ Ci/L. The principal radionuclides are isotopes of strontium, ruthenium, cesium, and cerium. Tritium is present in low concentrations. Long-lived alpha emitters may be present in trace amounts.

The presence of waste components in monitoring wells was detected on the basis of increasing total salt level, nitrate and sulfate ions at above-threshold levels, and tritium. Other radioactive nuclides were not detected in the wells, even those closest to the injection wells (200 to 300 m away). At the same time, gamma logging in injection wells revealed strong gamma fields (up to 6000 mR/h), indicating a rapid buildup of nuclides in the

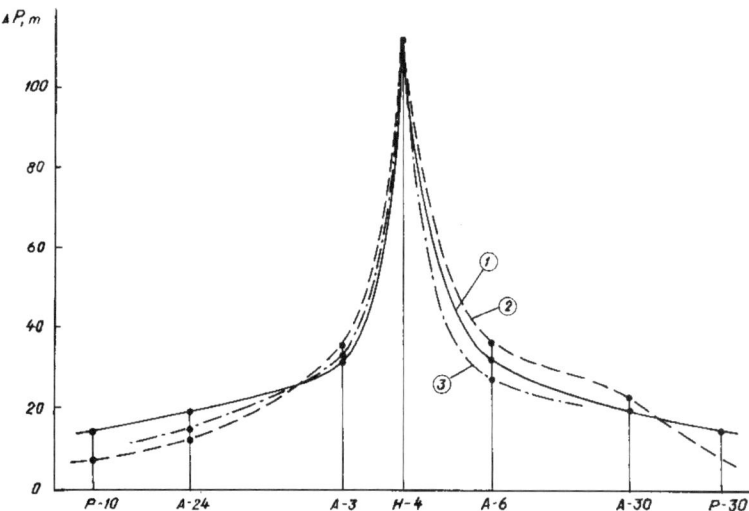

Figure 18. *Level of Subsurface Waters in Horizon II, Relative to Static Level, for Injection of Radioactive Wastes Into Well N-4 of the Area 18 Site. 1 - prediction, 2 - measured on April 23, 1967, 3 - measured on 16 March 1982.*

absorbing (lost-circulation) intervals of the reservoir. In geophysical (blind) wells located near the injection contour, the flow intervals of wastes injected during winter were identified by a drop in temperature. An analysis of measurements during operations at Area 18 showed that the capacity of the host formations was somewhat greater than had been supposed, because of the high effective porosity, which increased when wastes were in prolonged contact with the rocks.

Plotted in Figure 18 are combined waste dispersal contours (horizons II and III) for the operational period up to 1988 (injection into type N wells) and the outer contours for the operational period up to 1992 (injection into type AN wells). The outer contour of the wastes also includes "pillars" of subsurface waters between the N and AN contours. The contours were determined by simulations and calculations based on the presence of contaminating components in the wells, the volumes and regimes of injections, the subsurface flow structures, and so forth.

The total volume of wastes injected at Area 18, as of January 1, 1993, was 33.8 million m³, about 56% of the design volume. The area filled by wastes was 3.6 km², so that the total specific effective capacity of the horizon II and III host formations is on the order of 10 m. For a total effective thickness of 60 m for horizons II and III, this means that the mean effective porosity is 0.17. These figures are in good agreement with the data of Table 20.

Figure 19 is a map of the Area 18a (western) pilot disposal site for process wastes. This is essentially a development of the experimental site and is located nearby. The territory and some wells of the experimental site were included in Area 18a. The plan for this area included 16 injection wells and 42 monitoring wells (including those of the experimental disposal site). The monitoring network was later expanded: instrumented wells were drilled to horizon IV (buffer) around the operational injection wells.

The site complement includes four "batteries" of wells for pilot-scale batch disposal of wastes arbitrarily designated as high level. Each battery is composed of an injection well penetrating reservoir horizon II and a geophysical (blind) well in its immediate vicinity (14 to 20 m away). Geophysical measurements can thus be taken directly in the interval of the host formation without contacting the wastes. Instrumented (blind) strings were run along the axes of the first injection wells used to dispose of high-level wastes so that geophysical measurements could be taken. Additional injection and monitoring wells were later constructed in the northwest part of Area 18a, as planned, so that decantate generated by flushing and shutdown of the storage pools could be disposed of.

Wastes for deep injection disposal were prepared at the radio-chemical plant and then conveyed by pipeline to Area 18a. The design volume of intermediate-level wastes for injection was

The total volume of wastes injected at Area 18, as of January 1, 1993, was 33.8 million m³, about 56% of the design volume.

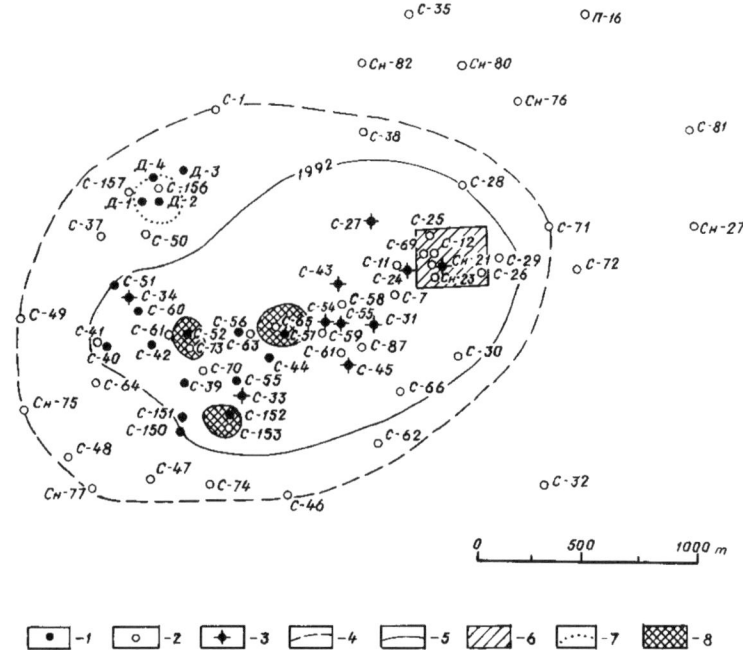

Figure 19. Location of Wells and Waste Dispersal Contours at Area 18a. 1 - injection wells, 2 - monitoring wells, 3 - shutdown wells, 4 - design waste dispersal contour, 5 - contour of intermediate-level wastes, 6 - experimental disposal site, 7 - contour of decantate (radioactive waste) from pools B-1 and B-2, 8 - contour of high-level wastes.

550 m³/d; the actual amount was far less and episodic, especially in recent years. The injection pressure did not exceed 1.6 MPa (up to 1.9 MPa in isolated cases). The pilot-scale disposal of arbitrarily designated high-level wastes was carried out in batch mode. Between one and three times a year, limited batches of 1000 to 2000 m³ of wastes were injected. Injection was performed in free-flow mode or under pressures of 0.5 to 0.6 MPa.

According to observations, the injection of intermediate-level wastes is marked by a regular increase in formation pressure in horizon II. The levels of the subsurface waters in horizons II and III of Area 18a are also seen to react to waste injection in Area 18. When the injection of process wastes is stopped, the liquid levels in the injection wells become established below the Earth's surface after 1 to 2 days or less, because of the relatively small volumes of the injected wastes and their high specific gravity in comparison with the subsurface waters. For the same reasons, the repressuring dome quickly evened out over the area of horizon II after injection ceases. The disposal of high-level wastes has almost no effect on the shape of the potentiometric surface of the subsurface waters in horizon II.

Intermediate-level wastes are represented by weakly alkaline solutions containing up to 300 g/L of salts. In chemical and radiochemical composition, they are similar to the solutions emplaced in the experimental site. High-level wastes have salt levels up to 220 g/L with a pH of 2 to 3 (see Sections 1.2 and 3.3). The preparation of high-level wastes for deep injection disposal includes adding complexing agents and correcting the acidity.

The disposal of high-level wastes is preceded by injection of weakly acidic solutions to prepare the host formation. After a batch of high-level wastes has been injected, it is displaced from the formation zone near the well screen. The mean specific activity of the mixture of wastes, conditioning solution, and displacing solution does not exceed a few curies per liter.

The presence of carbonate species in the host rocks and the interaction between acidic and alkaline high-level wastes leads to neutralization of the wastes and the formation of poorly soluble compounds that trap and coprecipitate the nuclides. Heating of the geologic medium together with radiation and chemical processes intensifies these interactions. According to laboratory studies, as much as 95 to 98% of the nuclides may enter the solid phase; the filtrate remaining in the pore space is an intermediate-level product.

The arrival of waste components near a monitoring well is detected through increases in salt level, content of nitrates, specific activity, gamma field values, and temperature. Figure 20 presents typical gamma and thermal logs from monitoring wells in Area 18a where intermediate-level wastes were registered. The observed and predicted dispersal contours for these wastes were plotted in Figure 19. The total activity of the emplaced high-level wastes was roughly a factor of two greater than that of the intermediate-level wastes, while the physical volume was about a factor of 30 less. Accordingly, the area of dispersal of the high-level wastes located in the intermediate-level waste zone was tens of times smaller.

Figure 21 presents findings on the dispersal of high-level wastes. Observations were carried out in an injection well and a nearby geophysical well. The gamma log records the filling of the horizon II permeable beds, with a total thickness increasing over time. The rising temperatures indicates that rocks were heated by energy released in radioactive decay. Waste localization in the permeable beds in the middle of horizon II was typical, as was the absence of intraformation leaks between beds, despite the long observation period. Thin clay beds within the host formation prevented the vertical redistribution of the wastes over the thickness of the horizon. The variation of the maximum temperature in the monitoring well (Figure 22) is in satisfactory agreement with the prediction. The temperature declined some time after injection ceased.

To gain a closer understanding of how the waste components interact with the host formation rocks, special tests were performed with decantate (intermediate-level waste) from storage pools.

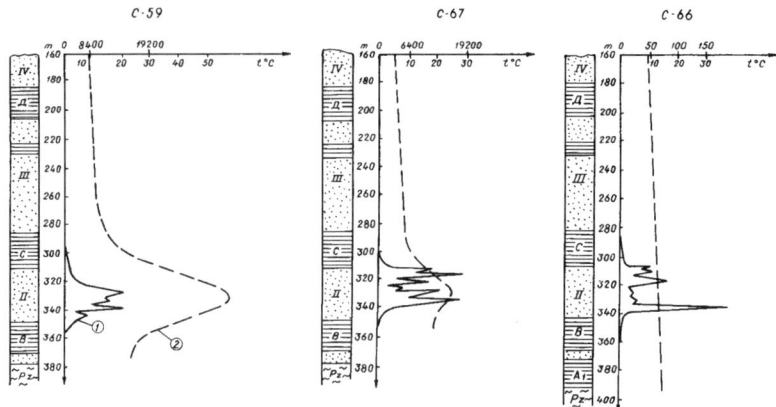

Figure 20. Logs Recorded in Monitoring Wells of the Area 18a Radioactive Waste Disposal Site. 1-gamma log, 2-temperature log.

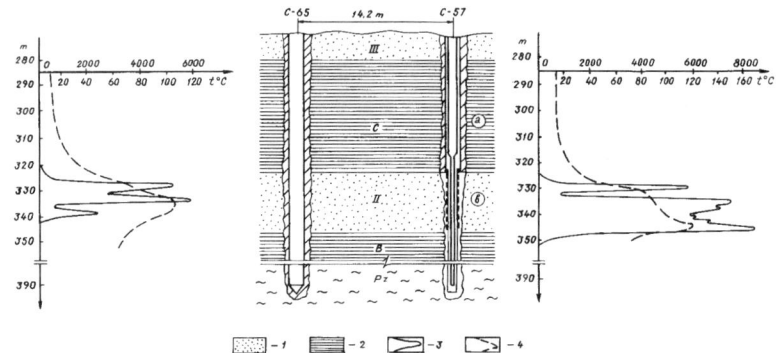

Figure 21. Geologic-Geophysical Section Along the Line Joining Wells S-65 and S-57. 1 - permeable sandy rocks, 2 - relatively impermeable clayey rocks, 3 - gamma log, 4 - temperature log (a) instrumental string, (b) injection string.

When the storage pools were closed, these wastes were injected, and subsurface water samples were taken from wells 50 m distant. The results of these tests were described in Section 3.3.

5.1.3 Environmental and Safety Aspects

Located in the immediate vicinity of the SCC disposal sites are surface storage pools for liquid radioactive wastes, slurries, and water; these can affect the subsurface waters. Areas 18 and 18a, and the area of the SCC as a whole, are located in a rapidly developing area in which subsurface waters are widely used for industrial and domestic purposes. Water-supply well fields 1 and 2 of Tomsk-7 (the city of Seversk) and of the SCC itself are located 8 to 9 km south and southwest of the disposal sites. The volumes of water extracted are no less than 20,000 to 30,000 m³/d.

The water supply well for the city of Tomsk (Levoberezhnyi well field) is in the land between the Ob' and Tom' rivers. Put in service in 1973, it produces over 140,000 m³/d. The eastern branch of this

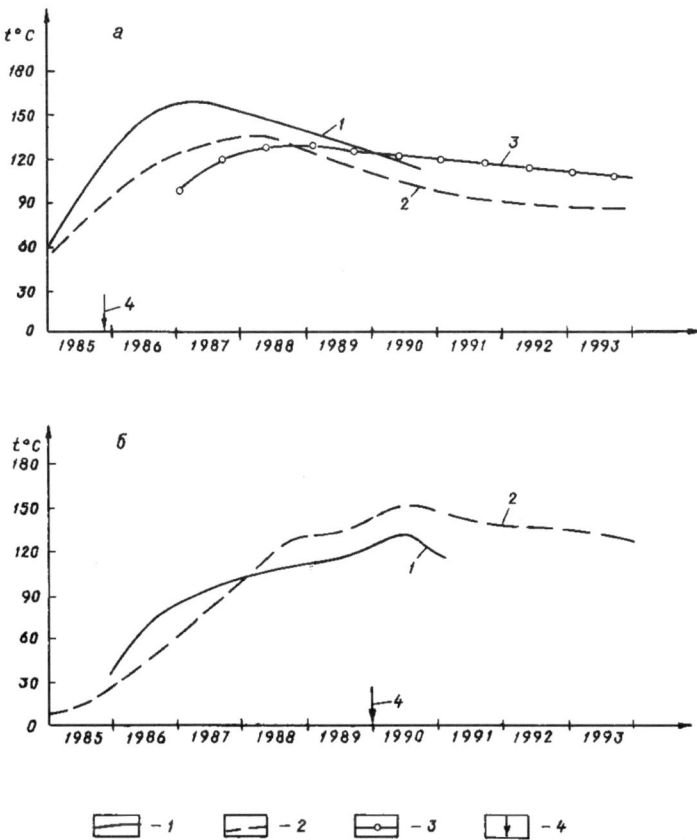

Figure 22. *Maximum Temperature of Horizon II, Area 18a, With Disposal of High-Level Wastes, (a) Wells S-57 and S-65, (b) Wells S-52 and S-73. 1 - temperature in injection wells, 2 - temperature in monitoring wells, 3 - predicted temperature, 4 - cessation of waste injection.*

well field, running along the left bank of the Tom' River, lies more than 14 km away from the disposal site (see Figure 23).

In the justification of deep injection disposal of liquid radioactive wastes in the early 1960s, the existence of well fields 1 and 2 was taken into account. The safety of operating these wells simultaneously with waste disposal was demonstrated. The presence of the disposal sites was considered when the Levoberezhnyi well field was established; it was brought into service 10 years after liquid radioactive waste disposal was begun. Well fields 1 and 2 and the Levoberezhnyi well field produce from aquifers V and VI, and in part IVa, of the upper aquifer complex. Relatively impermeable clay horizons E and D and buffer horizon IV separate these horizons from the lower aquifer complex, and from horizons II and III, which are used for disposal.

The subsurface well fields and the disposal sites are also adequately separated spatially. Well fields 1 and 2 and the Levoberezhnyi well field lie outside the subsurface exclusion zone of the disposal sites. It is predicted that waste components migrating in horizons II

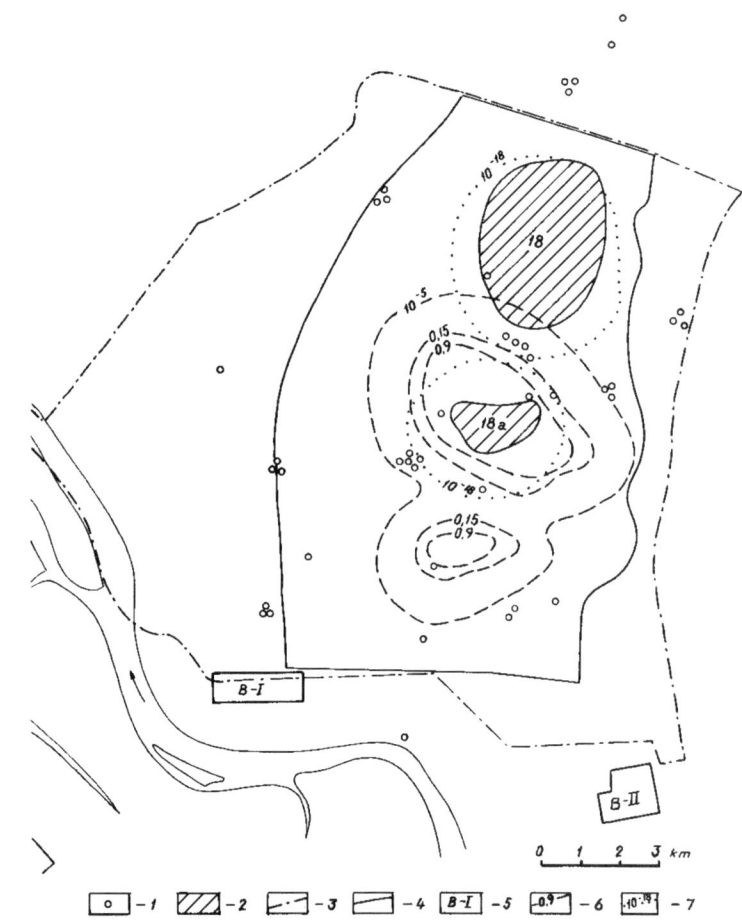

Figure 23. *Location of Wells Belonging to the Unified Subsurface Monitoring System, With Predicted Boundaries of Waste Migration. 1 - wells, 2 - Areas 18 and 18a, 3 - boundary of the plant's sanitary protection area, 4 - boundary of subsurface exclusion zone, 5 - wells withdrawing subsurface waters, 6 - lines of constant relative concentration of waste indicator for t-1000, 7 - similarly for strontium-90 (C = 10^{-19}) for t=1000 y.*

and III will not reach the well areas within 1000 years, even if waste interaction with the formation rocks is ignored. By way of example, Figure 23 shows the calculated migration of two radioactive waste components. The first is a stable indicator, such as sodium nitrate, that does not interact with the rocks. The second is ^{90}Sr, which is absorbed by the rocks (interphase distribution coefficient $K_p = 0.5$ cm^3/g). The predictive calculations were done by using the methods discussed in Section 3.4.

Analogous data were obtained for the migration of radioactive waste components in horizon IV should this horizon accidentally become contaminated. The migration times will be somewhat shorter for horizons V and VI if they are exploited by well fields 1 and 2, but physical and chemical interactions between waste components and rocks will prevent contamination of wells 1 and 2.

Because of the high hazard potential of the plants in the SCC, the waste disposal sites, and the surface radioactive waste storage facilities, and in view of the intensive use of subsurface regions in the SCC area, a "unified system of subsurface monitoring" has been established. The objectives of this system are to collect data about the condition of the geologic medium and subsurface waters, to study natural geologic phenomena and processes accompanying the deep injection disposal of wastes, and to define and delimit the areas affected by pollution sources. The unified subsurface monitoring system (USMS) includes monitoring wells, instrumentation and equipment, means of transportation, a special laboratory, and other facilities.

Monitoring wells inside and near the disposal sites belong to the internal monitoring subsystem of the USMS. These wells monitor all horizons identified in the section, including Paleozoic sediments, horizons II and III used for waste disposal, shallow horizons, and groundwaters. The location and number of these wells are chosen to simultaneously acquire data about waste dispersal in host formations and about the absence of contamination in shallower formations, as well as the development of hypothetical complications and accidents. Thus, an injection well is operated near a monitoring well that penetrates horizon IV (buffer formation); monitoring allows prompt detection of, for example, the preconditions for intercommunication between horizons via the borehole-casing annulus or other defects in the injection well, so that preventive action can be taken.

The external monitoring subsystem of the USMS includes nine batteries of wells along with single wells located outside the plant boundaries of the SCC. A battery has between three and five wells located quite close together and penetrating different horizons (Figure 23). The siting of external monitoring wells is determined by simulations and calculations of possible migration of waste components. Existing wells and the locations of well fields 1 and 2 are also taken into account; these wells can perform a monitoring role with respect to the Levoberezhnyi well field.

In these wells, hydrodynamic and geophysical observations and measurements are performed, and samples are acquired for various analyses (see Section 4.2). Between 20 and 25 species, including tritium, are determined in the subsurface water samples.

Within the USMS internal monitoring subsystem, observations are carried out by the disposal site operating team; external monitoring comes under organizations of Roskomnedra. Observational data establish the localization of wastes within the monitored boundaries of the disposal sites. No contamination has been observed beyond the boundaries of the sites and the subsurface exclusion zone. The locations of the waste contours in the section of the monitored region are illustrated in Figure 24.

Because of the high hazard potential of the plants in the SCC, the "unified system of subsurface monitoring" has been established.

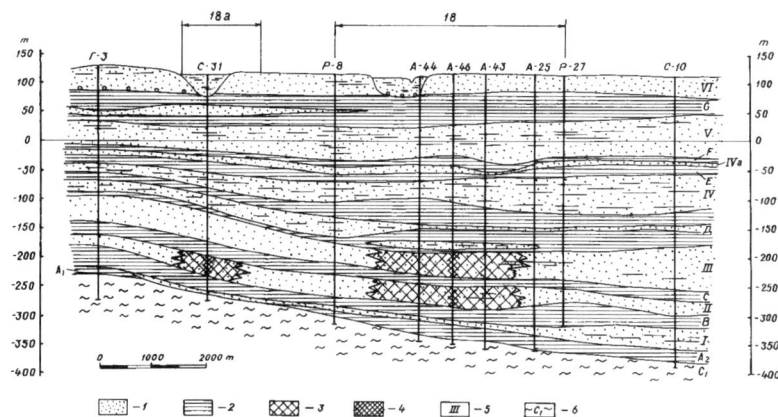

Figure 24. *Schematic Section in Areas 18 and 18a. 1 - permeable sand-clay rock, 2 - relatively impermeable clay rocks, 3 - region of waste disposal, 4 - areas of maximum concentration of nuclides, 5 - arbitrary nomenclature (symbols) of horizons, 6 - paleozoic sediments.*

Thirty years' experience in the deep injection disposal of radioactive wastes provides the grounds for a fairly objective operational safety analysis of Areas 18 and 18a. During this time, no events have been observed whose consequences have affected, or could later affect, the health-physics situation outside the boundaries of the disposal sites, sanitary protection areas, and subsurface exclusion zones.

The distribution of the wastes and the changes in the geologic medium, taken as a whole, have confirmed the theories formed earlier about the properties and structure of the geologic medium, as well as the predictions that were made. The high retention of nuclides by the rocks, as shown by experimental data, is an additional guarantee of the safety of liquid radioactive waste disposal.

Hypothetical complications (problems) and accidents were analyzed by methods discussed earlier (Section 3.5) for the conditions prevailing at the SCC. The results again showed that the consequences of accidents would not be catastrophic, and that the probability of accidents is extremely low.

Not surprisingly, the preliminary testing and the operation of the world's first deep injection disposal site for liquid radioactive wastes on a large scale was accompanied by certain difficulties and complications. Nevertheless, these complications did not suggest that preconditions existed for the uncontrolled development of negative processes that could cause serious accidents. The most significant complications that have occurred during the operation of the SCC disposal sites are examined in the paragraphs that follow.

When investigations were under way at the experimental site in 1963, a serious gas release took place in a monitoring well. The incident caused entrainment of gasified liquid containing radioactive waste components, which contaminated an area of less than 0.1 hectare. Special studies revealed that the gas evolution had been caused by the activity of denitrifying bacteria. These processes had been suppressed by the high-salt background of the wastes in the surface storage pool; they were activated in the zone of the host formation where wastes containing nitrates mixed with subsurface waters (zone of dispersion). The rapid discharge of gas from the well was aided by the formation of a "hanging" sand bridge (which prevented uniform outgassing of the formation), by the use of improper equipment to clear the sand bridge, and by the removal of the wellhead gate valve. After the method of waste conditioning had been corrected, rapid gas discharge from the wells ceased. The released gas was found to contain molecular nitrogen along with minor quantities of oxygen, carbon dioxide, and other species.

The condition of injection wells drilled in Area 18 at the beginning of the 1960s was found to be deteriorating. These wells had been in operation for 10 to 15 years and had been used to inject over 1 million m^3 of low-level wastes. Observations revealed that the pressure levels of subsurface waters in horizon IV (buffer) in a well close to the injection well were reacting to injection in horizon II and III (supposedly isolated). This effect was due to communication between horizons III and IV along the borehole-casing annulus of the injection well. A special investigation was instituted. Materials that were used to construct the well were examined, indicators (tracers) were injected, and geophysical measurements were taken. The results showed that the intercommunication had arisen through rapid withdrawal of subsurface waters (withdrawal was done to complete the filter zone during well construction) and the resulting entrainment of formation material. Evidently this had led to failure of the rock strata above horizon III and the formation of flow channels.

Since the interconnection between horizons was detected, the level in monitoring wells located near the injection wells is determined daily. The volume of solutions intruding into buffer horizon IV was insignificant, and it is predicted that contaminant migration in horizon IV will not reach the boundaries of the disposal site.

Later, all injection wells of this type were examined under a special program, and their capacities were downrated. The wells were then shut down in accordance with a special plan that included supplementary sealing of the borehole-casing annulus. Using a different technology, standby injection wells were put in service in 1988. Rock and groundwater samples were acquired when research and injection wells were drilled near existing injection wells. Examination of these samples confirmed the high retentive power of the sand-clay rocks for nuclides.

In 1963, a serious gas release took place in a monitoring well. Using a different technology, the standby injection wells were put in service in 1988.

145

Other difficulties included a leak ("blowout") of a box joint in the casing string of a well, leaks in piping and fittings, spills of contaminated samples, and loss of injectivity. None of these incidents led to contamination of objects outside the first zone of the sanitary protection area of the disposal sites (strict regime zone), nor did any incident create the preconditions for future contamination. They correspond to category 2 events in the international nuclear event scale (INES): "an event with potential consequences for safety," which is quite acceptable in view of the large volumes of wastes disposed of.

The environmental significance of deep injection disposal of liquid radioactive wastes at the SCC can be illustrated by comparing deep injection disposal and an alternative waste-handling technology. The experience of the Mayak enterprise in the southern Urals suggests that such a competing technology might be storage of wastes in many surface pools and basins as well as process vessels. Aerosol entrainment and surface contamination, the possibility of cyclones and other natural phenomena, and the location of a large city nearby would create a genuine threat and could not fail to have consequences for posterity.

5.2 Disposal Sites at the Mining and Chemical Complex

5.2.1 Geology

At the Mining and Chemical Complex (MCC), the waste disposal site is located in the "Severnyi" (Northern) area. The geologic conditions in that region have a number of features in common with Areas 18 and 18a at the SCC. The reservoir horizons are composed of weakly cemented sand-clay rocks and are isolated by relatively impermeable clay strata. The flow and capacity parameters are similar for the MCC and SCC reservoirs, as is the velocity of naturally moving subsurface waters. At the same time, there are significant differences because the Severnyi area does not have the same geologic structural plan, being located in the articulation zone of the West Siberian Basin and the Siberian Platform, the South Yenisei crystalline massif, and the southeast part of the Chulymo-Yenisei depression.

The site area is in an ancient erosional depression with a maximum depth of 550 m below the surface; the feature is filled with Jurassic sand-clay sediments (Figure 25). The thickness of the sand-clay strata decreases to the east, south, and southeast, and basement rocks crop out at the surface. The depression is bounded on the west by a fault striking north-south; the fault plane is clay-filled and isolates the rocks of the downthrown block from those of the upthrown block.

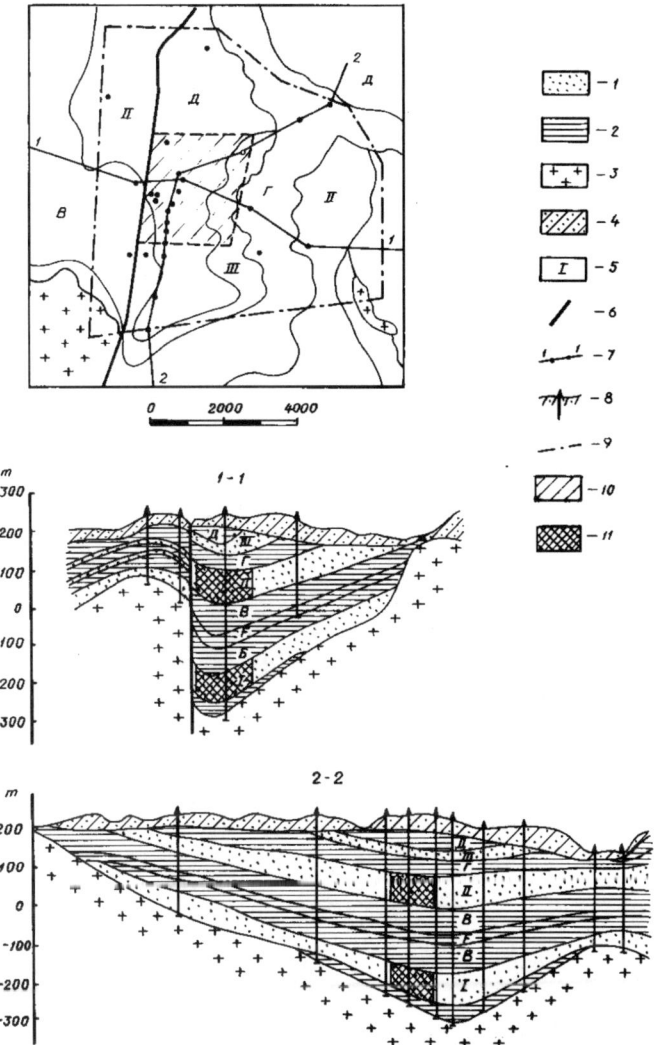

Figure 25. *Geologic Diagram of the Waste Injection Site. 1 - permeable rocks, 2 - relatively impermeable rocks, 3 - hard rocks, 4 - alluvial sediments, 5 - labeling of horizons, 6 - tectonic zone (barrier), 7 - line of geologic section and exploration wells (plan), 8 - exploration wells, 9 - boundary of subsurface exclusion zone, 10 - deep injection disposal site, 11 - areas of disposal sites.*

The special features of the depression thus include its clearly marked and self-contained synclinal character and the limited extent (in plan) of the sand-clay sedimentary formations filling it. As a result, favorable conditions for waste burial could be inferred as early as the preliminary stage of geologic exploration. The tectonic fault in the western part of the structure caused some concern. The geologic exploration included special studies of the fault zone and its hydrogeologic function. Wells were drilled to expose the fault plane, a geophysical study was done, and pumping tests were performed to record the piezometric levels in wells that had been drilled into the block on the other side of the fault. The natural seismicity rating of the region is less than 6.

It was found that the fault plane serves as a clay barrier both within the disposal site and in the area of possible influence of deep injection disposal. Evidence for this conclusion included the difference in the natural piezometric levels of subsurface waters in the downthrown and upthrown blocks (as much as 43 m) and the observations made during filtration tests. The finding was later confirmed by helium survey data and site operation results. The clay barrier is formed by clay beds lying at high inclination angles (i.e., steeply dipping), which were not disrupted but merely reworked by virtue of their high plasticity while the sand rocks were fractured and brecciated. Three to four kilometers north of the center of the disposal site was a region where horizons of the downthrown and upthrown blocks might communicate.

According to the geologic exploration, the floor and sides of the depression are composed of metamorphic gneisses, which are covered by variegated clays (residues of a Triassic-Jurassic weathered crust). The Jurassic formations are represented by alternating beds of sands and clays, within which are beds of permeable sand rocks with reservoir properties and beds of less permeable clay rocks having confining properties. Within the Jurassic formations, Lower Jurassic (Makarovskaya suite) and Middle Jurassic (Itatskaya suite) portions can be distinguished. The upper part of the section includes thin Quaternary formations. Figure 26 shows the stratigraphic column of the Severnyi disposal site.

The aquifer complex of the Jurassic sediments includes sand horizons numbered I, II, and III (from bottom up), which are separated by relatively impermeable clay horizons B, V, G, and D. Horizon I is underlain by relatively impermeable sediments of the weathered crust (horizon A).

Horizons I and II, located in the central part of the site at depths of 370 to 465 m and 180 to 280 m, respectively, were recommended as reservoir formations for waste disposal. Horizon II includes two strata of permeable rocks separated by a clay layer. The lower permeable stratum is used to dispose of low-level wastes; the upper acts as a buffer.

The reservoirs comprise medium-grained sands and weakly cemented sandstones and are characterized by the following composition: 70 to 80% quartz; 5 to 15% potassium or sodium feldspars, orthoclase, microcline, and plagioclase; 10% minerals of the mica and hydromica group; and 3 to 5% clay minerals. Within the reservoir horizons, subordinate beds of clay rocks are seen. The flow and capacity properties of horizons I and II used for waste disposal are described in Table 21.

The relatively impermeable horizons are made up of various clay rocks: argillite-like and "fat" clays (horizon B), aleurolites,

SYSTEM	SERIES	STAGE	SUITE	SUBSUITE	SYMBOL	ARBIRARY SYMBOL	LITHOLOGY	THICKNESS	DESCRIPTION OF ROCKS
JURASSIC	MIDDLE	BATHONIAN	Itatskaya	Upper	J it	D		20-50	Aleurolitic clays
						IIIa			Aleuritic sands
						D			Aleuritic argilite-like clays
						III		0-30	Arcosic sands
		BAJOCIAN	Middle Itatskaya		J it	G		30-50	Argillite-like carbonaceous clays
						II		50-95	Aleurites and aleurolites with interbedded sands, clays
									Carbonaceous clays
									Arcosic sands, locally with high content of clay and with interbedded clays
		AALENIAN	Lower Itatskaya		J it	V		40-75	Argillite-like clays with interbedded claycy sandstones
						F		0-24	Green argillite-like clays
	LOWER		MAKAROVSKAYA		J mk	B		30-75	Arkosic sands
									Gray argillite-like clays
						I		0-100	Gravely sands, breccia
									Unsorted fragments of rock with limestone cement
Rhaetian	UPPER	Rhaetian			Tr-J	A		0-43	Variegated kaolinic clays and breccias
Precambrian					pCm				Crystalline states or schists, gneisses

Figure 26. Stratigraphic Breakdown of Pre-Quaternary Sediments in the Region of the Mining and Chemical Complex

carbonaceous clays (in the upper part of the section), and interbedded clays with subordinate interbeds of sands, sometimes limy. The clay horizons are found throughout the site's area of possible influence, wedging out to the east and south at the sides of the depression.

Hydrogeologically, the region of the Severnyi disposal site can be regarded as a small artesian basin located in the downthrust block and open to the Chulym artesian basin on the north. Aquifer complexes in Quaternary and Jurassic sediments are known in the area, as is a complex of Precambrian metamorphic and igneous rocks. (The Quaternary formations and hard basement rocks are not well flooded and hold no interest for any practical use.)

Table 21. Flow and Capacity Parameters of Reservoir Formations in the Severnyi Site at the Mining and Chemical Complex

Parameter	Unit	Horizon I	Horizon II (lower part)
Depth	m	355-500	180-280
Thickness	m	55-85	25-45
Effective thickness	m	25-35	23-45
Total porosity	—	0.2-0.25	0.3
Effective porosity	—	0.07	0.08-0.12
Transmissivity	m^2/d	5-40	20-80
Hydraulic conductivity	m/d	0.3-1.6	0.1-2.2
Pressure conductivity factor	m^2/d	$1.6 \cdot 10^5$	$2.2 \cdot 10^5$
Pressure head over top	m	360-370	62-147

The principal aquifer horizons in Jurassic sediments are horizon I of the Lower Makarovskaya subsuite and horizon II of the Middle Itatskaya subsuite; these are composed of sands with clay interbeds. Horizon III is less flooded and does not occur throughout the region.

The recharge area of horizon I is located 7 km south of the Severnyi site; the main discharge area is the valley of the Kan River, 12 to 14 km from the site. The natural rate of movement of subsurface waters is 5 to 6 m/y, with the flow from south to north. The slope of the potentiometric surface is 0.003. The recharge region of horizon II is 4 to 5 km south of the site; this horizon discharges in the same area as horizon I and also partly into the valley of the Tel' River outside the area of influence of the waste disposal site.

The reservoir formations contain water with a low-salt content (under 0.3 g/L); the potentiometric heads over the roofs are 360 to 370 m for horizon I and 62 to 147 m for horizon II. The difference in the natural pressure heads of the subsurface waters is 2 to 3 m in the area of the disposal site.

According to special geophysical studies, horizons I and II feature a somewhat larger vertical flow inhomogeneity than the reservoir horizons at the SCC. But the intervals of maximum flow at the MCC cannot be identified in the various wells, this suggests more-or-less uniform waste dispersal in the reservoir formation.

The results of geologic exploration in the MCC region and the experimental disposal program at the SCC made it possible to recommend, with fair confidence, that a site for the deep injection disposal of radioactive wastes should be established. The following factors were taken into account:

- The reservoir horizons have an appreciable capacity and are isolated by clay beds from one another, from the overlying horizon III, and from the surface.

- The areal extent of the complex of sedimentary rocks that can be used for waste disposal is limited. There is a tectonic barrier on the west, which isolates the reservoir formations from the upthrown block and the permeable rocks in the direction of the Yenisei River valley.

- The velocity of naturally moving subsurface waters to the north and northeast is low. Sorptive holdup of nuclides by the rocks, given the low-salt content of the subsurface waters, limits waste dispersal to the previously established boundaries of the geologic medium.

- The synclinal character of the reservoir horizons means that wastes can be further localized in the most deeply buried portions of the structure, within and in the immediate vicinity of the disposal site. This effect is linked with the fact that the wastes are denser than the subsurface waters.

- The upper part of the geologic section in the disposal site area is not well flooded, so the use of horizon III for water supply purposes is not promising. Disposal operations present a minimal threat to natural resources of fresh subsurface waters.

- Geophysical and hydrodynamic investigations and an analysis of geologic and geophysical documentation from test wells have not revealed the signs of tectonic faults that would become channels linking the reservoirs with horizon III, surface waters or groundwaters. The barrier function of the tectonic fault (which later received the name "right bank") was established through special studies.

- Inferences about the geologic conditions in the disposal area are based on comprehensive and detailed investigations performed while a large number of wells were drilled (see Section 3.2); these materials are regarded as highly reliable.

In light of the above recommendation, we must consider that the limited areal extent of the reservoir horizons, I in particular, and the decrease in flow properties toward the sides of the structure impede the horizontal redistribution of formation pressures generated as a result of waste disposal. Therefore, fairly high injection pressures may be necessary; this will have a detrimental effect on the uniform filling of the reservoir formation with wastes.

To relieve the reservoir pressure, it was suggested that wells suitably distant from the injection wells be used to pump out

The wastes in the reservoir formations would disperse over no more than 6 km²
in the 30-year operational period and would then remain localized within the sanitary protection zone and the subsurface exclusion zone for at least 1000 years.

subsurface waters simultaneously with waste injection. Preliminary calculations and filtration tests at the geologic exploration stage confirmed that such a scheme was possible in principle. The optimal location and operating conditions for the injection and relief wells were determined at the justification and planning stage.

The volume of radioactive wastes to be disposed of was far smaller than the volumes at the SCC. There were grounds to expect that the wastes in the reservoir formations would disperse over no more than 6 km² in the 30-year operational period and would then remain localized within the sanitary protection zone and the subsurface exclusion zone for at least 1000 years. These circumstances were important in the decision to proceed with liquid radioactive waste disposal.

5.2.2 Waste Disposal Site

The Severnyi area used for deep injection disposal of liquid radioactive wastes at the MCC is located about 12 km from the main plant. Injection wells are arranged in linear fashion, the arrays for horizons I and II being combined. The distance between wells for each horizon is about 200 m. There are seven injection wells in horizon I and four in horizon II. The relief wells for horizon I are located 1 km away from the injection wells in a direction opposite (south) the natural movement of subsurface waters. The relief wells for horizon II are 1.5 km north. Because of the smaller volumes of injected nonprocess wastes, it turned out that these wells were not operated; they were used as observation wells (Figure 27).

Located between the rows of injection and relief wells are 46 observation wells for monitoring conditions in horizons I, II, and III. Another 42 observation wells are situated outside the disposal site proper. Exploration wells were included among the observation wells after inspection of their technical condition. Some of the observation wells are located in the immediate vicinity of injection wells. The disposal site is also equipped with standby injection wells; these were originally used as observation wells and were later refitted for injection service (prefix AN). By 1992, three of the standby injection wells had been placed in operation.

In accordance with the plan, intermediate-level process wastes have been injected into the horizon I reservoir since 1967. Injection rates have been up to 300 m³/d and injection pressures up to 1.2 MPa. Relief wells produce up to 300 m³/d. Results of hydrodynamic tests in the observation wells show uniform variations of the pressure heads associated with the injection of wastes and the withdrawal of subsurface waters. There is a repressuring dome in the region of the injection wells and a cone of depression in the region of the relief wells (Figure 28). Calculations show that the injection pressure would be as high as 5.0 MPa if the reservoir

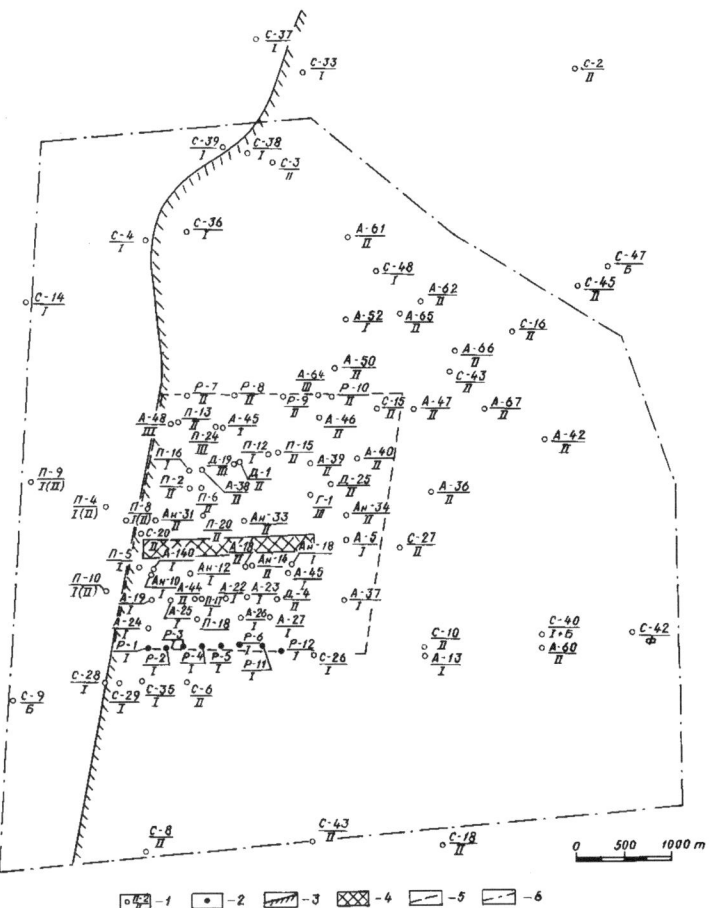

Figure 27. *Locations of Monitoring Wells in the Severnyi Disposal Site. 1 - operational wells, 2 - relief wells, 3 - tectonic barrier, 4 - injection contour, 5 - site boundary, 6 - boundary of subsurface exclusion zone.*

formation were not relieved. On the basis of an analysis of monitoring data, the operating regimes of the injection and relief wells and their interconnections were designed to achieve uniform filling of the horizon I reservoir.

The process wastes destined for disposal contained sodium salts, silica gel, and ions of several metals. The total salt content was up to 240 g/L, with a specific activity not exceeding 10^{-2} Ci/L. The radionuclide composition was similar to that of SCC wastes, including strontium, cesium, ruthenium, cerium, and several other short-lived nuclides.

The appearance of waste components in the area of the observation wells was detected by a rise in total salt level, an increase in the contents of nitrate ion, tritium, and specific activity (later), and a higher gamma level. These changes were seen in the most permeable intervals of the reservoir formation. Higher gamma values were also found below the floor of the horizon I reservoir in

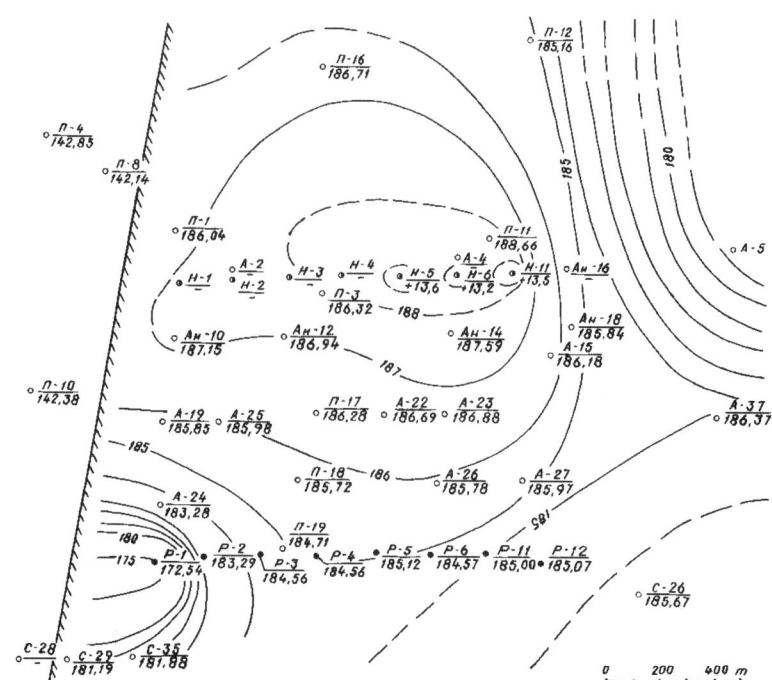

Figure 28. *Equipotential Line Plot for Horizon I of the Severnyi Disposal Site With Liquid Radioactive Wastes Being Injected in Wells N-5, N-6, and N-11, and Relief Well R-1 in Operation*

the sumps of the observation wells against relatively impermeable horizon A; such gamma anomalies coincide with the appearance of nitrates but sometimes precede increased fluid activity in the screen zone of the well (as in well P-11).

This phenomenon was explained by the transport of fine suspended particles of the pelite and clay constituents of rocks, containing nuclides in sorbed form, into the borehole by entrainment in formation liquid; these particles collected in the sump. Liquid samples taken from the screen zone of the well at the same time contained radioactive nuclides at levels below the acceptable DKB and threshold values. (Sediment from the samples was not analyzed because of its small volume.) The sediment that formed in the well over an extended period contained more nuclides than the downhole fluid. This phenomenon, seen in observation wells with screen zones close to the injection wells, allows us to suggest that nuclides migrate with the finely divided particles in the pore space of the reservoir formation. It is obvious that this kind of mass transport will play a significant role when the velocities of liquid movement are sufficiently high, that is, above critical values below which finely divided particles will remain in the rocks. Such conditions occur at deep injection waste disposal sites during injection of wastes and near the injection wells. The rate of mass transport will be lower than the movement of substances dissolved in the water.

The intervals in which the reservoir formation is filled by the wastes are also identified by the rise in temperature from energy released during radioactive decay, and by the temperature change that occurs when wastes are injected at temperatures different from the formation temperature. In some cases, there is also an effect due to intraformation leaks within the screen zones of the observation wells, between beds of the reservoir formation differing in permeability.

By way of example, Figure 29 is a plot of nitrate level and specific activity in the liquid from a screen zone as the interface between formation waters and wastes passed through the cross section of the well. This well is located 200 m from injection well N-6. The increase in the levels after 120,000 m^3 of wastes had been injected was caused by the passage of the waste dispersal zone.

Typical gamma logs and thermal logs for wells having wastes in the rocks of the filter zones are shown in Figure 30. Individual beds of permeable rocks in the reservoir formation are filled with wastes; this is clearest in geophysical ("blind") well A-4, whose interior does not communicate with the reservoir formation. The lower-permeability beds separating the permeable waste-filled beds do not contain waste components despite prolonged contact with them. There is no penetration of waste components into the relatively impermeable clay horizons overlying and underlying the reservoir formation.

Observations of the passage of the waste front through wells A-2, A-4, A-58, and P-3 were used to refine understanding of the filtration (flow) and dispersion characteristics of the reservoir horizons and plot their waste fillage. Waste dispersal in the bedded reservoir formation depends on the permeabilities of the beds. In turn, permeabilities are governed by the flow properties and the degree of colmatage (plugging) of the rocks by the clay mud during well drilling and by wastes during waste injection. The beds of highest permeability are filled first, and the waste dispersal fronts in these beds migrate faster than those in lower-permeability layers; this characteristic is clearly seen in the gamma logs of geophysical wells located near the injection well.

An analysis of geologic and monitoring data shows that these beds in the sand-clay sections of the MCC and SCC can be traced for tens to a few hundred meters. The beds then wedge out or are replaced by less permeable beds. A similar picture is seen for beds of reduced permeability; this on the whole leads to redistribution of the wastes over the thickness of the reservoir formation and an increase in the reservoir capacity.

However, plugging and lowered permeability occur in those intervals (beds) that absorb the most wastes. Solids suspended in the wastes are responsible for this effect, along with precipitation

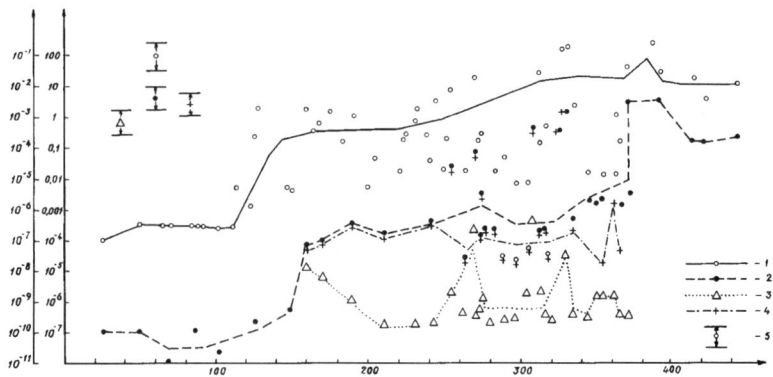

Figure 29. *Levels of Waste Components in Formation Liquid in the Well AN-14 of the Severnyi Disposal Site Versus Volume of Intermediate-Level Wastes Injected. 1 - NaNO₃, g/L, 2 - total activity, Ci/L, 3 - ⁹⁰Sr, 4 - ¹⁰⁶Ru, 5 - limits in wastes.*

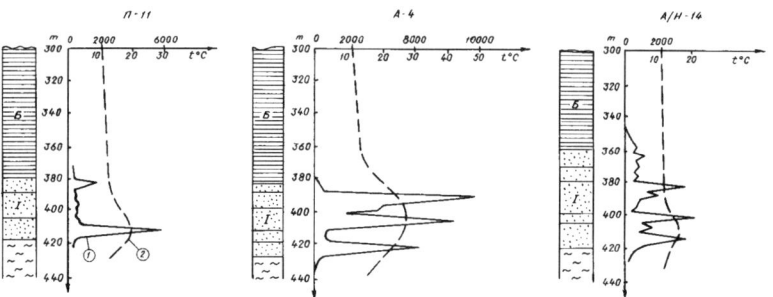

Figure 30. *Logs Recorded in Observation Wells at the Severnyi Liquid Radioactive Wastes Disposal Site. 1 - gamma log, 2 - thermal log.*

processes. The absorption of wastes increases in the intervals of the section that formerly accepted less wastes (and hence did not become plugged). As a result, waste distribution is equalized over the thickness of the reservoir formation, while the movement of the average waste front slows as the front moves farther from the injection wells. Some part of the permeable bed is not filled by the wastes. The total thickness of the reservoir formation filled by wastes is somewhat less than the effective thickness determined in geologic exploration. This does not, however, affect the specific capacity of the formation, because the effective porosity is large and increases after extended contact of the wastes with the rocks. (Wastes can disperse rapidly through thin zones if the injection pressures are high and the formation is hydrofractured.)

The waste dispersal contours in the reservoir formation are established by calculations and simulation. Monitoring data (pressure fields, component levels, and geophysical characteristics) enter into these calculations. Figure 31 shows a waste contour constructed on the basis of determinations of waste components, chiefly nitrates, in liquid samples taken from observation wells.

In 1974, it was noted that wastes were dispersing preferentially in the direction of well A-23. To smooth out the waste contour, the

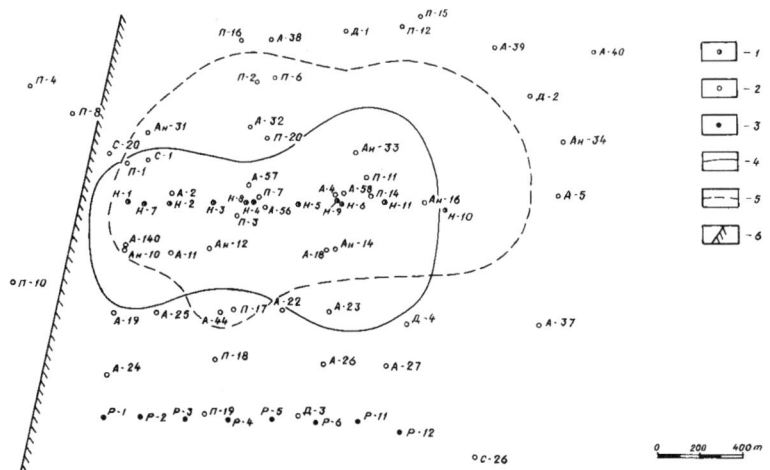

Figure 31. *Locations of Wells and Waste Dispersal Contours for the Severnyi Disposal Site. 1 - injection wells, 2 - observation wells, 3 - relief wells, 4 - contour of intermediate-level wastes based on nonradioactive components, 5 - contour of low-level wastes based on nonradioactive components, 6 - tectonic barrier.*

eastern relief wells were taken out of operation, and the western wells were activated; the waste contour evened out.

Waste dispersal in the reservoir formation corresponds, on the whole, to the forecasts. One indicator of this is tritium. Figure 32a,b shows the results of a measurement in which no elevated tritium concentrations are seen beyond the boundaries of the defined waste contour. The tritium monitoring technique was devised by L. I. Gedenov of the NPO Radium Institute. The composition of waters pumped from the relief wells is monitored, and the specific activity of these waters is under continuous surveillance.

Wastes of elevated activity (0.1 to 5 Ci/L), arbitrarily designated high-level wastes, have been disposed of at the Severnyi site since 1972. They are injected in batches one or two times a year (or less often); batch volume ranges from 1000 to 2000 m³.

The reservoir formation is prepared for high-level wastes by injecting weakly acidic solutions to decrease the buildup of nuclides in the zone near the injection well. Wastes are injected in a weakly acidic medium (pH 1 to 2) after pretreatment that includes conversion of poorly soluble compounds to soluble complexes.

Wastes are injected into two wells located in the common row of horizon I wells. High-level wastes are disposed of by free-flow injection or with minimal wellhead pressures. There is virtually no change in the potentiometric level of the subsurface waters in the well region.

Waste dispersal and heating processes in the reservoir formation are monitored by geophysical measurements from the instrumentation strings in the injection wells and also in "blind" well A-2, which is

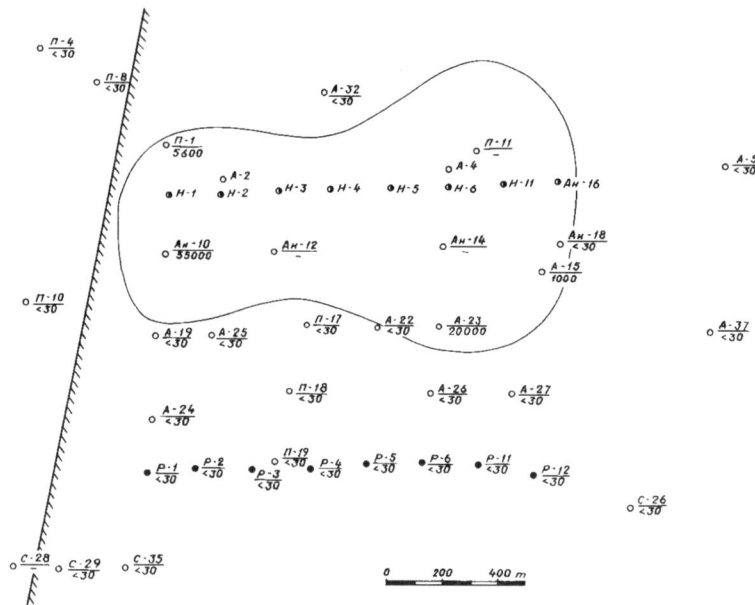

Figure 32(a). *Results of Tritium Determinations in Wells at the Severnyi Disposal Site, Horizon I. See Figure 31 for legend.*

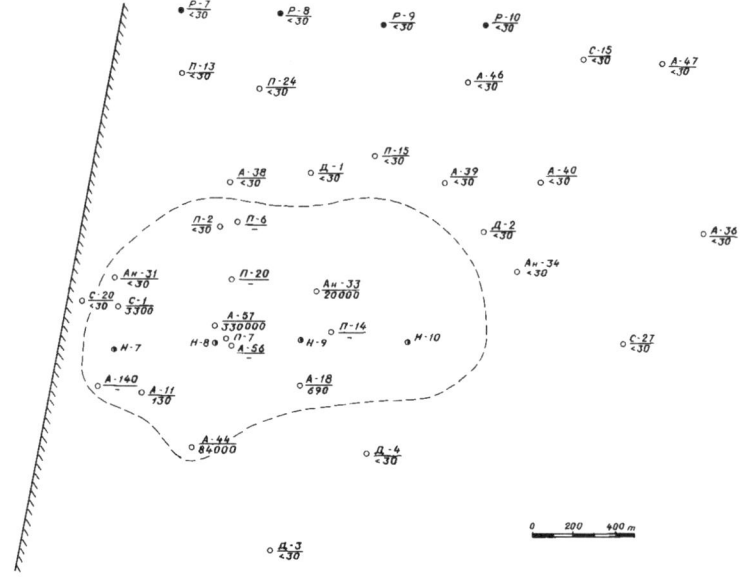

Figure 32(b). *Results of Tritium Determinations in Wells at the Severnyi Disposal Site, Horizon II. See Figure 31 and Figure 32(a) for legend.*

located 50 m from the injection well. Figure 33 presents results from geophysical monitoring. The permeable beds of the reservoir formation appear to be filled by wastes. The rock temperatures correspond, on the whole, to the forecasts.

As in the tests at the SCC, the disposal of high-level wastes is followed by a decrease in their acidity in the reservoir formation. This change results from the waste interaction with the carbonate

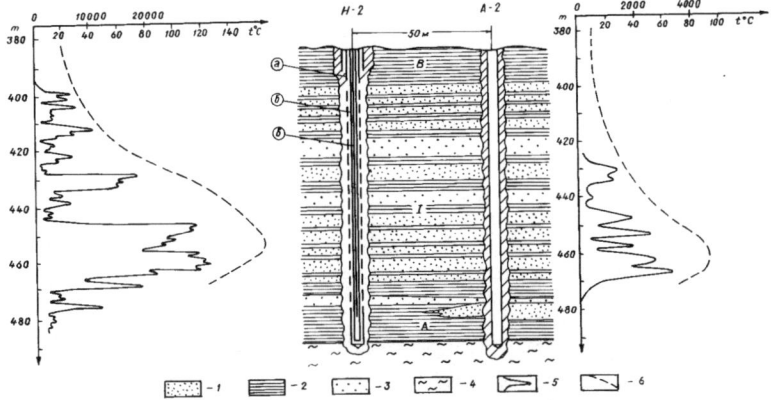

Figure 33. *Geologic-Geophysical Section Along the Line Joining Wells N-2 and A-2. 1 - permeable sand rocks, 2 - relatively impermeable clay rocks, 3 - sandstones, 4 - basement rocks, 5 - gamma log, 6 - thermal log (a) injection string, (b) tubing, (c) instrumented string.*

component of the rocks, radiochemical processes, elevated temperatures, movement of alkaline intermediate-level wastes with the filtrate, or contact with rocks holding intermediate-level wastes in the pore space. These processes cause the formation of poorly soluble compounds that trap and coprecipitate the nuclides. It is thought that 95 to 98% of the nuclides are converted to the solid phase and held in place. Observations provide some confirmation of this notion. The dispersal area of the high-level wastes is several orders of magnitude smaller than that of the intermediate-level wastes, even though the high-level wastes contain the greater part of the nuclides.

Low-level nonprocess wastes have been disposed of in the horizon II reservoir of the Severnyi site since 1968. The injection rate is as high as 600 m³/d, and pressures range up to 2.0 MPa. The reservoir formation is not relieved, since the pressures are equalized over fairly short times. The waste injection is accompanied by a regular rise in levels in horizon II.

The nonprocess wastes contain salts and detergents. The salt content is up to 10 g/L. The radionuclide composition is generally similar to that of the process wastes. The presence of wastes in the region of an injection well is detected by an increase in the total salt level of the formation fluid (nitrate and sulfate ions). The dispersal of radioactive nuclides is markedly slower than that of the chemical pollutants, except where wastes are injected at high pressures and flow rates (see below). In well A-57, which is situated 85 m from the injection well, the content of nonradioactive components is close to that in the wastes; radionuclide levels are two to three orders of magnitude less.

Figure 31 shows a scheme of waste-component dispersal in the horizon II reservoir formation according to 1992 data. The waste contours were determined by monitoring and calculation. The

The dispersal area of the high-level wastes is several orders of magnitude smaller than that of the intermediate-level wastes, even though the high-level wastes contain the greater part of the nuclides.

The presence of the "right bank" fault quite close to the injection wells and their repressuring domes, together with the fact that the aquifer complexes of the upthrown block contact the Yenisei River Valley to the west, meant that vigorous monitoring was needed in the fault area to ensure the safety of the disposal operation and prevent environmental pollution.

total volume of low-level nonprocess wastes injected into horizon II was 2 million m^3. Figure 32 (b) gives results of tritium monitoring in the wells working horizon II, and Figure 34 is a schematic section through the disposal site region showing the waste dispersal contours.

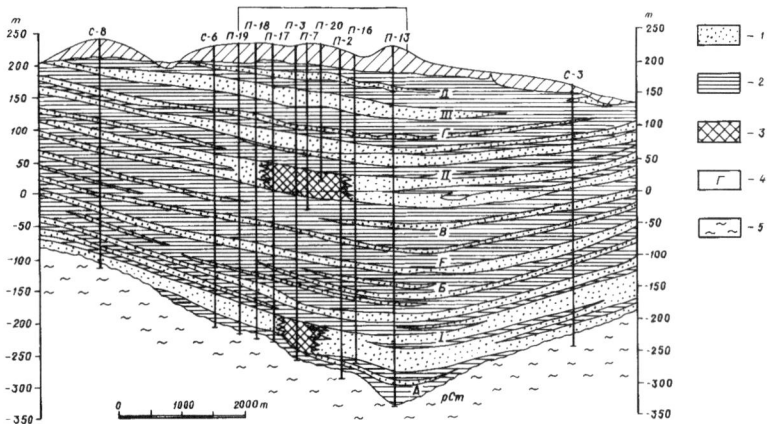

Figure 34. *Schematic Section Through the Severnyi Site. 1 - permeable sand-clay rocks, 2 - relatively impermeable clay rocks, 3 - waste disposal zone, 4 - regions of maximum nuclide concentration, 5 - labeling of horizons, 6 - Precambrian sediments.*

5.2.3 Environmental and Safety Aspects

The deep injection disposal of liquid radioactive wastes at the MCC differs from that at the SCC because there are 1) no other water consumers in the area of possible influence of the site or in the immediate vicinity, and 2) the shallow aquifers hold little promise as sources of water supply. These facts simplify the monitoring of subsurface waters. Nevertheless, a network of monitoring wells was created within and outside the boundaries of the disposal areas at the Severnyi site (see Figure 27).

The presence of the "right bank" fault quite close to the injection wells and their repressuring domes, together with the fact that the aquifer complexes of the upthrown block contact the Yenisei River valley to the west, meant that vigorous monitoring was needed in the fault area to ensure the safety of the disposal operation and prevent environmental pollution.

As was mentioned in the previous sections, the geologic exploration of the region included special studies of the fault, which established its barrier function in the area of possible influence of the burial site. Observation of the potentiometric surface of the subsurface waters during disposal operations (in wells P-1, P-4, P-5, P-10, S-1, and S-20) confirmed that the fault plays an isolating role. The newly advanced hypothesis of recent tectonic processes, as well as the danger of induced seismicity in the fault zones of the Severnyi site, were considered as the Special Design Office of the Russian Academy of Sciences Institute of Earth Physics went to work on the problem. The institute devised a continuously functioning seismic monitoring system comprising KAGK-D-1 and POISK apparatus. The system includes 12 peripheral measurement stations; up to three components of the seismic field are logged at each station. The data cover an augmented frequency band. A central station is equipped to record signals in digital and analog form, and cables provide the lines of communication. Figure 35 shows the layout of the peripheral stations. Observations over a period of years did not reveal any signs of induced seismicity due to the injection of wastes, even though seismic events linked with distant earthquakes were noted several times.

Highly accurate geodetic observations were performed to identify possible geodynamic phenomena. The locations of benchmarks and leveling lines are also shown in Figure 35. The data show a change in the position of the surface near the injection contour (at the threshold of detection) when the horizon II injection wells are operating. Changes in the operating conditions of horizon I are not reflected in the surface position.

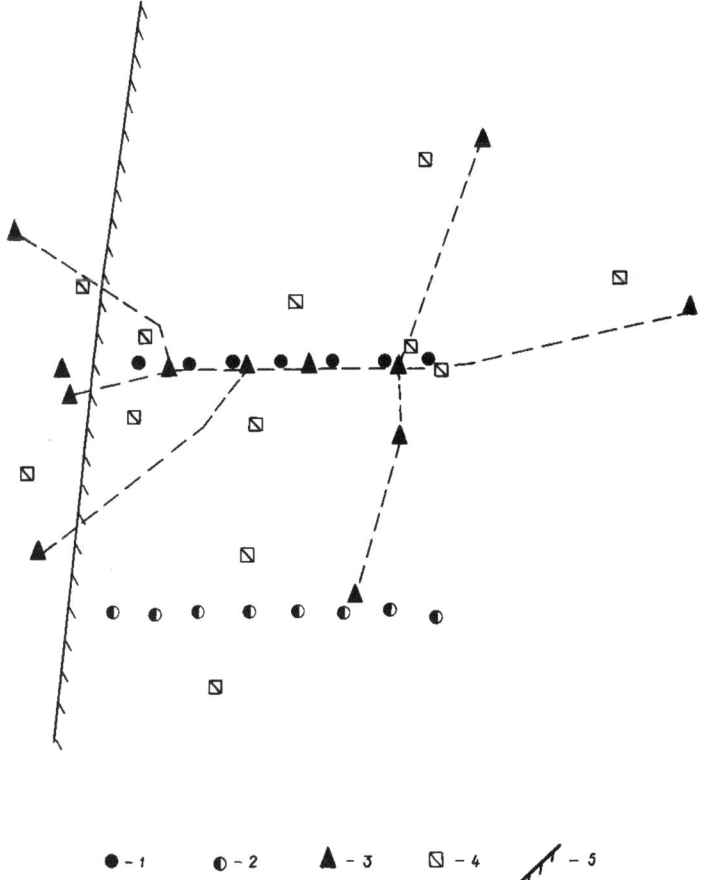

Figure 35. *Layout of Seismic/Geologic Monitoring Stations. 1 - injection well, 2 - relief well, 3 - leveling line and bench mark, 4 - seismic monitoring station, 5 - tectonic barrier.*

The operation of disposal sites has created difficulties and complications, though none were in the nature of accidents.

The operation of disposal sites has created certain difficulties and complications, though none were in the nature of accidents. The most typical problem is leaking surface equipment and piping at fittings, shutoff valves, and unions. These leaks were identified promptly, and design features have made containment possible.

The absorbing capacity of the wells deteriorated because the filtration zones of the reservoir formation became plugged when alkaline process wastes were injected. A special method devised by the Russian Academy of Sciences Institute of Physical Chemistry was used to restore the injectivity. In well N-2, the bottom of the lift string had become closed by formation material entrained into the well ("sand plug"). The string was shortened by about 4 m, and the well became serviceable again.

Rapid dispersal of wastes, exceeding the forecast rate, was seen in the reservoir of horizon II. This problem must be examined more closely. As an economy move during initial operation of horizon II, wastes were injected in one or two wells at elevated flow rates and injection pressures. Analysis of downhole liquid samples and gamma logging revealed waste components in observation wells near the injection wells.

In 1973, waste components including radionuclides were detected in a remote observation well (A-38). The species were found in a thin interval of permeable rocks identified within horizon II on the basis of gamma log data. Well A-38 was located inside the boundaries of the disposal site, but this scale of waste dispersal contradicted the theories held about waste behavior and nuclide retention by rocks.

Investigation showed that the phenomenon resulted from waste dispersal through a zone of elevated permeability that had formed because of the high injection pressures (near, or even exceeding, the hydrofracturing pressure of the formation). Because of the drastic increase in the pore space in the flow interval, sorption processes were weakened so that the nuclides could propagate over long distances.

Temperature anomalies signaling vertical leaks within horizon II were noted in remote observation wells P-20, P-13, and P-24. Flowmeter data confirmed flow between the interval where maximum flow resulted from elevated injection pressures and permeable beds located in the same horizon but receiving little or no injected material.

As recommended, the injection pressures were lowered and more wells were put in service. These actions ensured uniform filling of the reservoir formation by wastes, flow of the wastes in the porous medium, and retention of the nuclides by rocks. The concentrations of waste components and the gamma values in well

A-38 fell to background levels. The contents of waste components also declined in other wells. The results of observations in well A-57 show effective retention of nuclides by the rocks.

The safety analysis of deep injection radioactive waste disposal at the Severnyi site included predictions of waste migration in the reservoir formation after the shutdown of the site. Analog simulation and elementary finite-difference calculations were used at first; later, digital computers were used. The methodology of these predictive calculations was discussed in Section 3.4.

Figure 36 shows the predicted contours of the distributions of a waste indicator and a radionuclide in horizon I. (The indicator is a stable species that does not interact with the rocks; the radionuclide here is ^{90}Sr.) The calculations allowed for hydraulic dispersion, radioactive decay, and the distribution of flow properties in plan. The results of these predictive calculations were used to establish the boundaries of the subsurface exclusion zone.

The calculations to predict the migration of waste components took account of the difference in density and viscosity between the wastes and the subsurface waters, the synclinal nature of the structure, and hydraulic dispersion. A comparison with the solution of the two-dimensional problem shows that density effects reduce waste dispersal for the central part (the "core") of the waste contour (see Figure 37).

An analysis of hypothetical complications and accidents showed that the consequences of possible accidents do not affect the radiation situation or lead to contamination of groundwaters outside the sanitary protection zones and subsurface exclusion zones. Among the scenarios considered was the appearance of flow paths between the downthrown and upthrown blocks at the fault barrier. The first sign of such an event would be a change in the subsurface water levels in the wells on the upthrown side. The breakthrough time and amount of contaminants appearing in the upthrown block would depend on the position of the waste contour as controlled by naturally moving subsurface waters in aquifers on the downthrown side. For example, if a flow path appeared in a portion of the fault barrier south of the contour of northward-migrating contaminated waters, the migration of this contour could be retarded; pure waters or waters containing nuclides formerly transferred to the solid (rock) phase and now leached out again could flow into the upthrown block.

The contamination of aquifers on the upthrown side can result from a combination of unfavorable factors: the coincidence of a flow "window" and the contamination contour. The development of clay rocks in the fault zone, with their greater sorption capacities, will prevent the penetration of radioactive contaminants into the upthrown block.

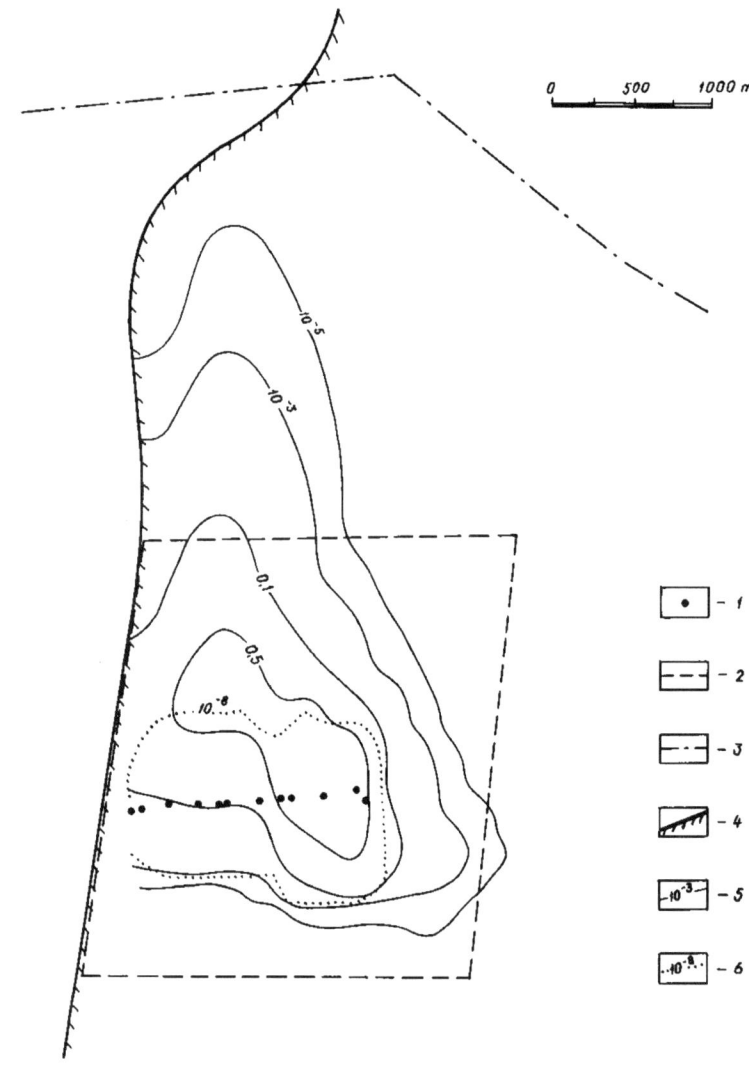

Figure 36. *Predicted Migration of Radioactive Waste Components at the Severnyi Disposal Site. 1 - injection well, 2 - boundary of disposal site, 3 - boundary of subsurface exclusion zone, 4 - tectonic barrier, 5 - relative content of waste indicator at t = 650 y, 6-relative content of ^{90}Sr at t = 950 y (K_p = 0.6 cm^3/g).*

The sand-clay sediments of the upthrown block are also included in the subsurface exclusion zone, whose boundaries are placed about 1 km west of the disposal site. The time required for contaminated waters to reach this boundary, without allowance for contaminant retention by the rocks, is estimated to be 100 to 200 years. If the time required to develop a flow "window" and to complete physical and chemical interactions is taken into consideration, the time scale becomes hundreds to thousands of years, during which the radioactive nuclides will decay.

Nevertheless, monitoring of the fault barrier must continue throughout the operational and postshutdown phases of the disposal site. The chief monitoring methods are hydrodynamic,

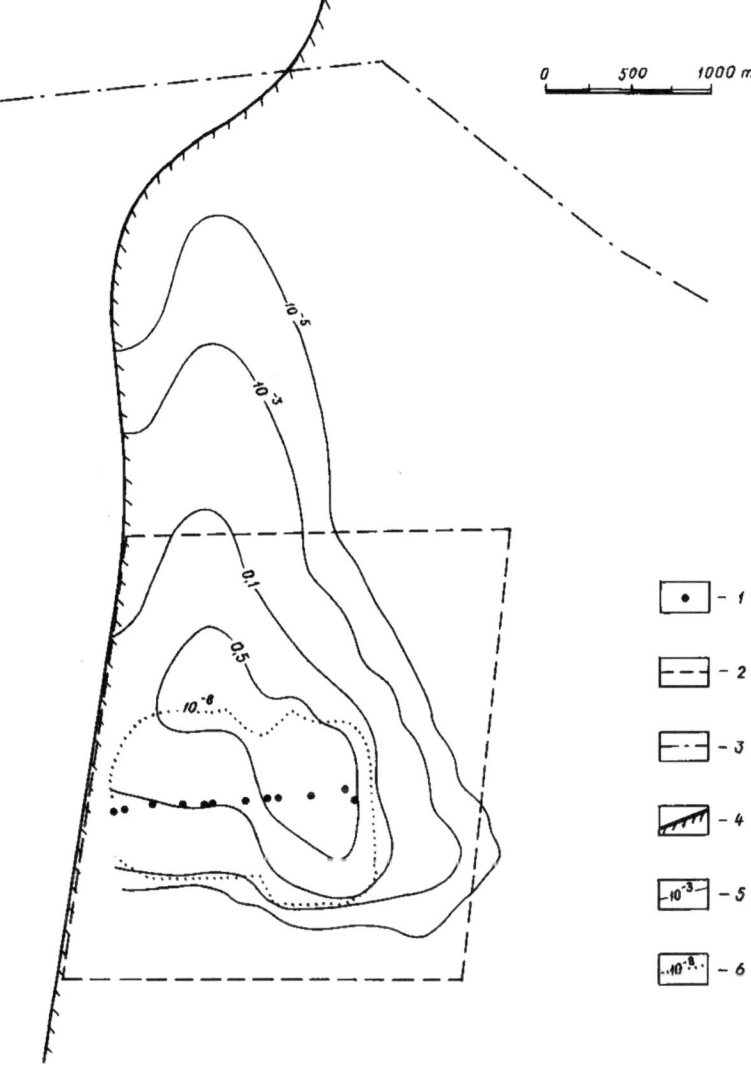

Figure 37. *Predicted Migration of Radioactive Waste Components at the Severnyi Disposal Site. Different density (specific gravity) of the waste and underground water. See Figure 36 for legend.*

hydrogeochemical, and geophysical observations in wells located in both the downthrown and upthrown blocks. For accident response purposes, it is desirable to work out a technology for restricting the migration of contaminants.

The deep injection disposal of liquid radioactive wastes at the Severnyi site of the MCC has made it possible for an appreciable fraction of the wastes to be isolated from the sphere of human habitation. As a result, the need for potentially dangerous surface storage facilities and ponds has been obviated: a very important point in preventing the consequences of radiation on the environment, especially considering that the MCC facilities are close to the Yenisei River.

The deep injection disposal of liquid radioactive wastes at the Severnyi site of the MCC has made it possible for an appreciable fraction of the wastes to be isolated from the sphere of human habitation—a very important point considering that the MCC facilities are close to the Yenisei River.

5.3 Disposal Site at the Scientific Research Institute of Nuclear Reactors

5.3.1 Geology

The geology of the NIIAR site is distinguished by the widespread occurrence of sedimentary rocks up to 2300 m thick. Aquifer complexes containing mineralized waters (including brines) have been identified in the geologic section along with horizons of relatively impermeable rocks with confining qualities. The subsurface waters display vertical hydrogeochemical zonation indicating that the deep-lying aquifers are isolated from the surface and from subsurface fresh waters. As a consequence, geologic exploration here was directed primarily at studying the deep portions of the section, lying beneath the Vereian formation (Middle Carboniferous), which is widely encountered on the Russian Platform and serves as the confining formation for many oil fields including those of the Volga region at Ul'yanovsk.

The early results of the geologic exploration showed this region to be highly promising for the deep injection disposal of liquid radioactive wastes. Geologists at Saratov University later arrived at a similar conclusion, namely that the region holds promise for the deep injection disposal of a variety of industrial wastes.

A key consideration in identifying deep horizons as suitable for radioactive waste disposal is the high-salt content in the waters saturating the formations. If the toxicity of natural brines and wastes is assessed by the ratio of their principal components to the DKB values (maximum permissible concentrations) in potable water, the relative toxicity figures are of the same order of magnitude. In addition, the toxicity of radioactive wastes will decline over time as their radionuclides decay.

Seven permeable zones were identified in the sedimentary section. Those considered promising for waste disposal are found in permeable zones III and IV (Lower Carboniferous), which lie at depths of 1440 to 1500 m and 1130 to 1410 m, respectively. These are overlain by the Vereian horizon. Permeable zones V and VI are located above the Vereian; these also contain mineralized waters and fulfill the role of a buffer horizon. Thick, relatively impermeable Permian sediments separate zones V and VI from the shallow freshwater horizons. Figure 38 illustrates the stratigraphy of the region's sedimentary formations.

Permeable zone III is of terrigenous rocks of the Yasnopolyanskii horizon, which is part of the Tul'skii stage (Lower Carboniferous). This zone is represented by weakly cemented clayey sandstones with intercalated clays and aleurolites enriched in carbonaceous matter. At the boundary between zone III and the overlying sediments, there is a thin bed of vuggy sandstones with limy cement.

Eratem	System	Stage	Symbol	Arbitrary Symbol	Litolo-gy	Thickness	Descriptions of Rocks (see next page)
Cenozoic	Quaternary			VII		50	Sands, loams, clay
Mesozoic	Triassic Jurassic			B		75	Clays, interbedded sands, aleurolites
Paleozoic	Permian	Tatarian				135	Dense brown clays, argillite-like, interbeds of aleurolites and sandstones
		Kazanian		VI		120	Clayey marls, limestones and dolomites with interbeds of gypsum
		Sakmarian				95	Halite rocks gypsumsand anhydrites
	Carboniferous	Gzhelian-Kasimovskii		V		260	Limestones and dolomites, light gray, organogenic-detrital. Vuggy-porous, jointed
		Moscovian				315	Limestones and
		Vereian		A		45	Argilites and clays
		Bashkirian		IV		40	Limestones
		Okaian-Serpukhovskii				265	Dolomitized limestones and dolomites, wealy cemented sandstones with thin interbeds of clays
		Tul'skii		III		90	Sandstones with interbedded clays and argillites
		Tournaisian				90	Alternation of dark gray shales/slates/schists, clayey limestones and argillites, silicate-bituminous limestones

Figure 38. Stratigraphy of Paleozoic and Mezo-Cenozoic Sediments in the Vicinity of the NIIAR

Permeable zone IV is composed of carbonate rocks of the Oksko-Serpukhovskii and Bashkirian horizons, which consist of fissured limestones (some dolomitized) and dolomites. Table 22 gives the flow and capacity characteristics of permeable zones III and IV.

The waters saturating permeable zones III and IV are of chloride-sodium type with a total salt content of 205 to 260 g/L. The waters also contain calcium, magnesium, potassium, sulfate and chloride ions, and trace components. The natural rate and direction of movement of the subsurface waters are hard to evaluate; the rate may be no more than centimeters or tens of centimeters per year, and the direction may have varied from one geologic epoch to another.

Tectonically, the region lies in the Ul'yanovsk interblock zone, the western margin of the Melekesskaya depression, near its boundary with the Stavropol'sk depression. The structural plan of the crystalline basement clearly displays a block structure, which is reflected in small-amplitude deformations of flexure type in the overlying sedimentary strata. According to several researchers using geophysical data and well-logging results, there are no fault-like features in sediments of Carboniferous or greater age in the NIIAR region, nor do the prerequisites exist for such faults.

Table 22. Flow and Capacity Parameters of Reservoir
Formations at the NIIAR

Parameter	Unit	Permeable Zone III	Permeable Zone IV
Depth	m	1440-1550	1130-1410
Thickness	m	45	320
Effective thickness	m	35	80
Total porosity	-	0.14	0.01-0.26
Effective porosity	-	0.06	0.02
Transmissivity	m²/d	35	6
Hydraulic conductivity	m/d	1.0 (brine)	0.1
Pressure conductivity factor	m²/d	5×10^5	4×10^4
Pressure head over top	m	1585-1650	1250-1280
Salinity	g/L	230-260	205-220

The Ul'yanovsko-Mokshenskaya zone of fault dislocations lies in crystalline basement rocks 6 to 8 km south of the disposal site. Typically, oil fields are found in the continuation of the zone in Lower Carboniferous formations; this indicates that the geologic formations have fair isolating properties even in zones of regional tectonic faults that can be traced in the basement and the sedimentary formations lying below the confining formations.

According to seismic exploration, the most uplifted portions of the disposal site proper are located in the northwest, while the greatest subsidence is seen in the southeast. Permian sediments in the center of the area display slight local uplifting with a southeast strike: a structural "nose" with an amplitude of 20 to 30 m and width on the order of 1 km. According to information on the most recent tectonics, the NIIAR region lies in a low-amplitude structural zone and is characterized by a reduced density of megajointing.

The main reservoir horizon that was first selected was permeable zone III, composed of sand-clay sediments. Sand reservoirs are recommended for the disposal of liquid radioactive wastes at the SCC. It was suggested that the dominance of block (primary) porosity over fracture (secondary) porosity ensures uniform filling of the reservoir by wastes. But the interaction of the low-salt (sometimes ultrafresh, i.e., distillate) wastes with the brine-saturated sand-clay rocks complicated the injection of wastes, as will be discussed. Zone IV was subsequently chosen as the main horizon for this reason.

The formations selected as reservoirs had the needed capacity and flow parameters (Table 22); they were isolated from shallower horizons by confining formations; and they lay in the zone of impeded water circulation. Because there was no doubt about the suitability of the site for safe disposal of liquid radioactive wastes by deep injection, construction and operation of a pilot plant (well R-3) was begun even before drilling of some exploration wells had been completed or the entire test site had been erected.

5.3.2 Waste Disposal Site

The test site for the disposal of nonprocess wastes originating in NIIAR research facilities is located within the site boundary and close to the building where wastes are pretreated before deep injection disposal. The site includes five injection wells penetrating permeable zones III and IV and 32 observation or monitoring wells (Figure 39).

The wastes handled at the site include solutions from decontamination of equipment, plant compartments and protective garments; shower water from entryways; process water discharges; and water from fuel storage pools. They contain phosphates, oxalates, sulfo acids, nitrates, and oils. The dry residue is up to 3.6 g/L, the specific activity up to 1×10^{-5} Ci/L. Nuclides include isotopes of cesium, strontium, ruthenium, zirconium, and rare-earth elements. Long-lived nuclides are present in trace concentrations.

The test facility was placed in service in 1966 with injection in well R-3. High-level wastes were injected at a rate of up to 500 m^3/d and at pressures up to 5.0 MPa.

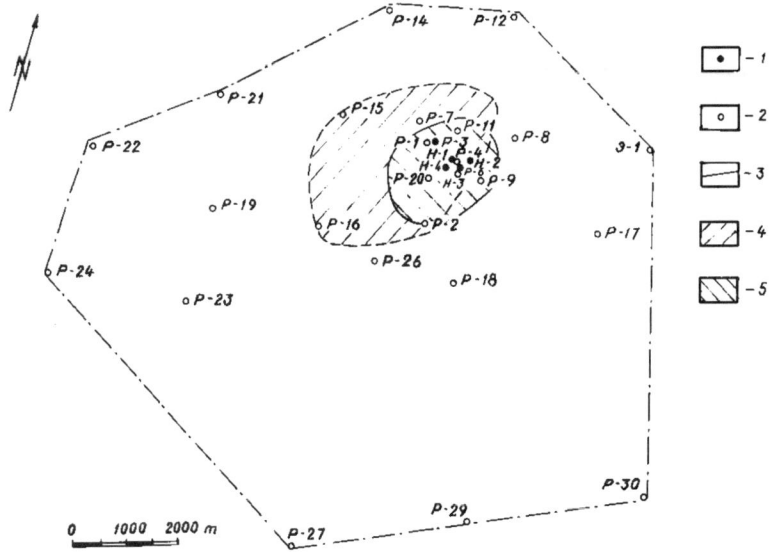

Figure 39. *Locations of Wells and Waste Dispersal Contours in Reservoir Formations of the NIIAR Disposal Site. 1 - injection wells, 2 - monitoring wells, 3 - waste contour in permeable zone IV, 4 - waste contour in permeable zone III, 5 - boundary of subsurface exclusion zone.*

Formation water (brine) was injected to test the experimental plant. Based on test results, the parameters found for the reservoir horizon were close to those determined by geologic exploration. In another test, fresh water with a macrocomposition resembling that of the NIIAR wastes was injected. The injectivity of the well dropped sharply (pressure rose as flow rate declined). In order to prepare the filter zone of the formation, hydro-fracturing was carried out by a method used in oil wells (three-step procedure). The injectivity of the well now improved and it became possible to operate well R-3 in the requisite mode.

In 1969, gamma logging in monitoring wells R-10 and N-1 revealed elevated gamma field values. Analysis of the geologic-geophysical documentation from these wells showed that waste dispersal over sizable distances was related to the filling of a thin (about 5.0 m), permeable formation of limy, vuggy sandstones found in the roof of permeable zone III.

The preferential filling of a vuggy sandstone formation by the wastes was accounted for by the sharp decrease in permeability of the clayey sandstones in zone III when the brine in the pore space was replaced by fresh waters or wastes with low-salt levels (distillate). Similar behavior has been seen in the development of oil fields and was attributed to hydration of the clay component of the rocks. Hydrofracturing enhanced the permeability of the thin bed of vuggy sandstones.

There was also another explanation of waste dispersal over appreciable distances: lighter wastes "float" to the roof of the reservoir formation by density-driven convection (gravitational segregation).[37] This hypothesis, however, is opposed because of 1) the clear confinement of the flow interval to a thin bed of vuggy sandstones, as shown by gamma log data; 2) the results of hydrofracturing; and 3) the observed dispersal of wastes in permeable zone IV.

The waste contour in zone III did not go beyond the sanitary protection zone, but further operations may allow the wastes to prematurely reach the design boundaries. For this reason, it was recommended that permeable zone IV be used as the main reservoir formation for the experimental disposal site. Zone III was put on standby and can be used if needed for the disposal of highly saline wastes. The total volume of wastes injected into permeable zone III was 0.6 million m^3; into zone IV, 1.5 million m^3.

Zone IV is composed of carbonate rocks with both primary and secondary porosity. The waste dispersal in this zone is much less given the large injected volume. Figure 39 shows contours of the wastes in zones III and IV; these are based on both monitoring data and predictive calculations.

Waste injection is accompanied by a regular increase in formation pressures. The lowering of the subsurface waters in the monitoring wells results from both the change in formation pressures and the lower salt content (and hence the reduced density) of the liquid in the borehole as waste filtrate is admitted. This fact makes the hydrodynamic data difficult to interpret. Measurements with a downhole pressure gauge should be performed as part of the monitoring program.

The appearance of waste components near a monitoring well is detected chiefly by gamma logging, for example, a rise in gamma field values as compared to rock intervals that contain a waste filtrate. The identification of radionuclides and other waste components in liquid samples taken from the screen zones of monitoring wells typically comes long after their presence has been revealed by gamma logging, which deals with the distribution of nuclides in the rocks outside the well casing.

Several factors are responsible for this:

- The dispersal, with the waste filtrate, of the main gamma-emitting component ^{137}Cs, as well as ^{106}Ru, and ^{144}Ce, through the permeable zones of the reservoir formation, which have predominantly fracture porosity and have relatively little power to retain nuclides on their brine-saturated carbonate rocks.

- The buildup of nuclides in the rocks near the well, because these rocks contain the filtrate of the clay-type mud used in drilling the well.

- The dilution of the low-density waste filtrate when it is admitted to the interior of a well originally filled with natural brine.

Figure 40 presents typical gamma logs from the monitoring wells of the experimental disposal site. The wells are located at various distances from the injection wells. Wastes filled individual permeable zones as distinguished by elevated gamma field values throughout the thickness of permeable zone IV and as seen by the formation of a hydraulic dispersion zone (displacement of wastes and formation waters). The latter process is due to the nonuniformity of the rocks and the time variation in the injectivity profile of the injection wells, which in turn results from plugging of the permeable intervals.

Figure 41 is a plot of (relative) gamma field strength over time. The measurement was taken in the interval of permeable rocks in the middle of zone IV. The log, recorded in well R-9, covers the period when the cross section of the dispersion zone passed through the well. Unit field strength is taken equal to the maximum value measured in 1990. This corresponds to an (^{137}Cs) activity of about 1×10^{-5} Ci/dm^3, close to the activity of the wastes.

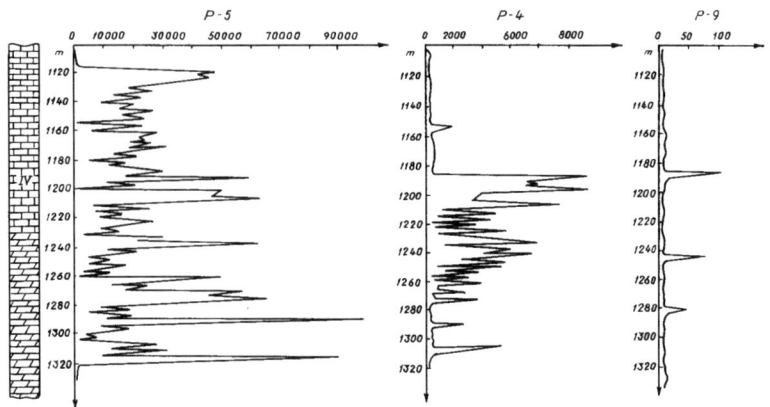

Figure 40. *Gamma Logs From Monitoring Wells at the NIIAR Disposal Site*

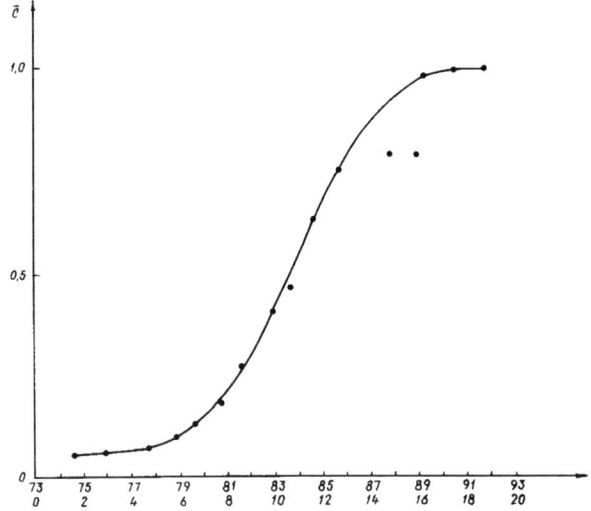

Figure 41. *Relative Contents of Radionuclides in Rocks in the Reservoir Zone Near Well R-9*

The dispersion zone near well R-9 was 1 to 1.5 km wide. In the dispersion zone, the density of the waste/formation brine mixture changed from the density of the brine to that of the wastes; accordingly, the horizontal density gradient is small. As a result, density convection (gravitational segregation) has much less of an effect on waste dispersal. The alternation of zones with high-permeability (fractured) rocks and low-permeability rocks (chiefly with primary porosity) over the thickness of zone IV causes the flow properties to be anisotropic, thus diminishing the effect of density convection. This follows from the geophysical data illustrated in Figure 39. In spite of the extended observation period (including downhole monitoring), zone IV, in which the continuous string is cased, shows no signs that the wastes are "floating."

Observations of waste dispersal in permeable zone III reveal a nonuniformity in plan: waste components were observed in the very distant well R-16 but not in R-8. It was suggested that this might be due to decreased depth of the formations toward well R-16 (slope 0.005) and the presence of the southeast-striking structural nose (positive structure), along whose axis the permeability might be higher.

The predominantly fractured character of the porosity in zone IV made possible the disposal of limited batches of wastes with an elevated solids content (up to 1 g/L suspended solids) formed in the sumps of the process vessels when the wastes were pretreated for injection. Before disposing of such wastes, the residue in the vessels was dispersed, and the wastes were injected into the reservoir formation at the highest possible rate to ensure the liquid in the screen zone of the formation would move at a rate higher than critical velocity (sedimentation rate of the suspended solids). Injection of these solids-laden wastes was followed by injection of solids-free solutions or water. The injection pressure did not exceed 5.0 MPa.

5.3.3 Environmental and Safety Aspects

The disposal site includes 32 monitoring wells used to check on the condition of the geologic medium. These wells are distributed in the region of the injection contour and as far as 10 km away from the site center (Figure 39). The monitoring wells expose permeable zones III and IV used for waste disposal, and also the shallower horizons, including the quite shallow freshwater horizons. Hydrodynamic and geophysical measurements are performed in these wells, and subsurface waters are sampled for a variety of analyses.

A deep injection exclusion zone (reserve) is set aside in the site area (Figure 39). A well producing subsurface waters from shallow horizons is located 4 to 5 km southwest of the boundary of the exclusion zone, while subsurface waters are withdrawn for curative use from the Upper Carboniferous sediments to the northeast of the exclusion boundaries.

Several hypothetical complications and accidents during waste disposal operations were analyzed, including leaks in the casings, deterioration of the cement seal in the annulus, and destruction of surface equipment (see Section 3.5). The results showed that such events would be detected promptly by the existing monitoring service before they are fully developed, and their consequences would not cause contamination of subsurface waters beyond the subsurface exclusion and sanitary protection zone boundaries. The development of complications can be forestalled by special practices including preventive actions. With regard to negative impact and probability of occurrence, geologic factors rank last among all possible problems. Monitoring during the operation of the experimental disposal site has made it possible to refine the capacities of the zone IV reservoir.

Zone IV has an effective thickness of 50 m, an effective porosity of 0.02, a specific capacity on the order of 1 m, and a dispersion zone length of around 1.5 km. Calculations of waste dispersal components show the following: After waste disposal in permeable zone IV, dispersion and flow nonuniformity over a 28-year period (up to year 2000) will yield a waste contour with a radius of 2.5 to 3 km from the site center. After shutdown, waste dispersal within the defined contour will be greatly retarded as nuclides are transferred from the pore liquid to the rocks. After some 300 years, radioactive decay will lower the activity of nuclides in the pore liquid and rocks to values below the thresholds for classification of a liquid or solid as a radioactive waste.

Several small deposits of oil have been discovered in the Volga region at Ul'yanovsk in recent years. Calculations were, therefore, done to predict the effect of crude oil production on the deep injection disposal of wastes at the NIIAR site. The Zimnitskoe, Filippovskoe, Allagulovskoe, and Zapadno-Labitovskoe oil fields are multilayer fields confined to domes of reef origin. They are located 12 to 25 km from the site boundaries. Oil pools have been found at depths of 900 to 1500 m in Tournaisian, Yasnopolyanskii, Bashkirian, and Vereian sediments (Carboniferous). No oil-bearing formations have been observed above the Vereian horizon; this indicates the significance of the Vereian as a regional roof for underlying rock complexes. The crude oils are mostly in the heavy class.

Predictive calculations have been done for the nearest of the fields, the Allagulovskoe, which is 19 km from the center of the disposal site and 12 km from the southern boundary of the subsurface exclusion zone. The basis for the calculation was a crude production of 500,000 m^3/y from the Bashkirian horizon (Middle Carboniferous, permeable zone IV). The radius of the influx boundary of subsurface waters will be around 2.8 km for a 25-year production life of the oil field and around 5.6 km for a 100-year production life. These values are much less than the distance to the disposal site.

The most frequent problems in operating the experimental (pilot) disposal site were failures of surface equipment and leaks at unions. These problems were promptly detected and remedied. In one instance, the casing of a monitoring well leaked; the leak was detected and repaired. The areal dispersal of wastes in a thin interval of the roof of permeable zone III (well R-3), discussed earlier, may be considered a complication.

The most significant problems, however, were groundless accusations directed against NIIAR and the deep injection disposal operation by irresponsible journalists, politicians, and a few scientists. These people have built their reputations by compiling slanders (some quite amusing) about nuclear industry enterprises and speculations on environmental problems. Following are two such accounts, and many others could be cited.

In 1990, for example, seismic surveying near the NIIAR was done using the "common depth point" method. The work was performed by a geophysical crew from an organization of the Ministry of Geology. Two shots were performed in which 300 to 600 kg of chemical explosive was detonated in shallow (15 to 20 m) boreholes quite close to the residential zone of the NIIAR in Dimitrovgrad. The explosions were approved in January 1990 by the agencies having jurisdiction and were carried out in June without further advance notice.

The populace took the blasts for earthquakes, and the explosive gases ejected from the well for a UFO. The local press, and later the national newspaper *Izvestiya* (August 17, 1990), published articles saying that the "earthquakes" were connected with the deep injection disposal of wastes at NIIAR. On this basis, a group of scientists at the Russian Academy's Institute of Earth Physics (M. B. Gokhberg, I. G. Reisner, N. V. Shebalin, and V. V. Shteinberg) went to the government predicting a major environmental catastrophe in the Middle Volga region. The administration of the Ul'yanovsk oblast demanded that activities at NIIAR be stopped.

A government commission headed by Academician V. N. Strakhov identified the real causes of the "nuclear earthquakes" and assessed the safety of deep injection disposal operations at the experimental site. The local press carried stories giving a correct account of events,[63, 64] but *Izvestiya* could not bring itself to do so.

In 1992-93, staff members at Kazan' University claimed that the pilot disposal site at NIIAR was situated in the zone of a deep (abyssal) fault and that the fault plane was providing a route for radionuclide contamination of mud sediments in the Cheremshanskii embayment of the Kuibyshev reservoir. The location of the site in the fault zone turned out to be based on a 1:25,000,000 map (1 cm to 25 km), on which it is difficult to tell that the NIIAR site is actually 6 to 8 km from the fault (the Ul'yanovsko-Mokshinskaya dislocation zone). The mud samples had been taken at a place where supposedly clean discharges from NIIAR had been dumped in the 1960s; the discharges had contained radioactive nuclides that accumulated in the muds.

The results of disposal operations at the NIIAR site have confirmed that the use of deep carbonate sedimentary formations for the disposal of industrial wastes is promising. Some features of deep injection waste dispersal have been clarified, particularly the way in which the fracture pore space is filled and the nature of the

The primary benefits of deep injection disposal at NIIAR are: 1) that more than 2 million m³ of wastes containing radioactive substances has been isolated from the sphere of human habitation, and 2) that these wastes have not been discharged into the waters of the Kuibyshev reservoir.

dispersion phenomena. It has been found—contrary to established opinion—that density-driven convection has little effect on the waste dispersal.

The primary benefits of deep injection radioactive waste disposal at NIIAR, however, are 1) that more than 2 million m^3 of wastes containing radioactive substances has been isolated from the sphere of human habitation, and 2) that these wastes have not been discharged into the waters of the Kuibyshev reservoir. The accumulation of radioactive wastes in special facilities at the surface, which do not afford any guarantee that groundwaters will be safeguarded from radioactive contamination, has been prevented.

Studies and experiments have established that deep injection disposal at the NIIAR site is also a possibility for other wastes from nuclear plants and from the Dimitrovgrad industrial region. Such wastes include those from electrochemical plants, such as the Dimitrovgrad automotive equipment factory, and from nuclear power plants (after multieffect evaporation). The disposal of such wastes will not affect the safety of disposal operations or the size of the subsurface exclusion zone.

5.4 Basic Scientific and Practical Conclusions

The results of 1) investigations to justify and create deep injection disposal sites for liquid radioactive wastes, 2) experience in disposal site operations, and 3) monitoring waste conditions in reservoir formations have made it possible to state the following general conclusions, which will be of interest for the comprehensive use of subsurface resources and the solution of various hydrogeologic problems:

1. It is possible and technically feasible to place (dispose of) various industrial wastes, including liquid radioactive wastes, in deep aquifers on an industrial scale, provided the formations meet certain requirements. Deep injection disposal ensures the localization of wastes within defined subsurface boundaries. The burial of wastes is not accompanied by detrimental impacts on the environment outside the bounds of the sanitary protection zone and the subsurface exclusion zone.

2. Porous, permeable rocks of sand-clay or carbonate composition, lying in beds underlain and overlain by relatively impermeable rocks with confining qualities, can be used for deep injection waste disposal. The geology and hydrogeology of proposed sites must be studied to determine whether the sites comply with the requirement of localizing wastes in well defined subsurface regions, and also to acquire starting data needed for justification and design.

3. The methods and equipment used in geologic exploration and investigation make it possible to obtain data sufficient for establishing disposal sites. Predictive calculations of deep injection disposal processes are, on the whole, confirmed by factual data collected during 30 years' operation of existing disposal sites.

4. In investigating proposed disposal sites, the following characteristics are most important: capacity and flow properties of the rocks, flow nonuniformity (both horizontal and vertical), relatively impermeable rocks overlying the reservoir formations, isolation of the reservoirs from shallower horizons, potentiometric level of subsurface waters, nature and velocity of the natural movement of subsurface waters, interaction between the wastes and the geologic medium, and geologic-technical conditions of well drilling and construction. The process of studying, justifying, and predicting burial processes is much more difficult for reservoir horizons that include karst voids, zones of anomalously high permeability, or tectonic structures of fault type.

5. Reservoir horizons usable for the deep injection disposal of industrial wastes are bedded systems. Sand-clay reservoirs are composed of beds and interbeds varying in permeability and traceable in area over limited distances. Carbonate reservoirs primarily consist of horizontal zones of increased permeability (resulting from fracture porosity), zones in which block (primary) porosity predominates, and relatively impermeable rocks.

When wastes are injected into a bedded reservoir formation, the beds and zones of higher permeability are filled first. Plugging of the absorbing beds leads to redistribution of the absorbing intervals and the incorporation of lower-permeability beds, as well as redistribution of wastes between beds, all resulting in wastes filling the majority of permeable beds in the reservoir formation.

A zone of hydraulic dispersion develops at the interface of wastes and formation waters. This, together with the bedded structure of the formation, reduces the effect of the density difference between wastes and subsurface waters (density-driven convection) on waste dispersal. Over an extended period of time, density convection occurs in the central parts of the waste contour when the beds are tilted (synclinally or anticlinally), and the density of the wastes is quite different from that of the subsurface waters. When wastes are in prolonged contact with rocks, the wastes fill blind and capillary pores as well as "block" (primary) porosity. The effective porosity increases and approaches the total porosity.

6. The scale of waste dispersal is affected by the injection pressure. When injection takes place in the hydrofracturing regime, wastes can move rapidly through thin zones formed in the formation. For controlled maximum injection pressures below the hydrofracturing values, the change in the pressure-head regime in aquifers is from 1 to 15% of the initial natural heads. This change does not lead to appreciable geodynamic processes.

7. Physical and chemical processes in the system comprising wastes, rocks, and formation waters cause radioactive nuclides and some chemical compounds to transfer to the solid phase, as a sorbate on the rocks and on precipitates of insoluble substances. These processes occur most quickly when the subsurface waters, the wastes, or the waste filtrate have a low-salt content. If the salt background is high, the predominant process by which nuclides are transformed to the solid phase is precipitation in the system comprising wastes and subsurface waters. The flow of subsurface waters desorbs only part of the nuclides and some compounds. The transport of the remaining nuclides into the pore liquid is retarded and resembles leaching.

   As a result, the nuclide dispersal is delayed relative to the aqueous phase, and nuclides build up in the region of the injection contours. It is possible, in principle, to speed up the transfer of radioactive nuclides from pore liquid to the solid phase by using special technologies to pretreat the wastes, pretreat the reservoir rocks, or carry out injection.

8. Relatively impermeable horizons (chiefly clay) overlying reservoir formations provide isolation for the wastes in the reservoirs. Thirty years of observations have not yet established the flow of wastes and the diffusional transport of nuclides on appreciable scales. Thin clay interbeds within reservoir horizons split the flows of wastes into shallower and deeper beds of permeable rocks; thus acting as local confining beds.

9. Fault planes that have become so filled with clay that (according to geologic exploration data) a hydraulic barrier is formed ensure the isolation of disposal-relevant reservoir horizons located to the sides of the fault plane. This has been confirmed by extended observations. The flexure-like form of clay beds of relatively impermeable horizons at faults ("benches") in the crystalline basement does not mean a loss of isolating power, provided the horizon does not contain regions where the clay has a high sand content (flow "windows").

10. The processes of deep injection disposal can be monitored. The use of hydrodynamic, hydrogeochemical, and geophysical techniques makes it possible to gain an overall view of the waste distribution and the processes that are parallel to disposal. Monitoring data can be used to optimize the injection conditions and demonstrate safety.

11. The most critical subsystem in a deep injection disposal system comprises the injection wells, which largely determine the efficiency and safety of disposal operations. Along with the construction technology, the wells must have special design features based on the geologic conditions of drilling, reinforcement, and completion, and the characteristics of the subsurface waters and wastes. All well construction work must be monitored, and quality must be evaluated. The main focus must be on the distribution and quality of the cement in the borehole-casing and tubing-casing annuli and on the tightness of casing strings, unions, and welded joints. In completing and operating a well, it is desirable to avoid drastic actions on the casing string, the cement, and the rock around the well; such actions might include repeated perforation, explosions, the application of high injection pressures (higher than the formation hydrofracturing pressure), rapid withdrawal of water with entrained formation material, and enlargement of the filter zone diameter in a sand-clay section.

12. Analysis of actual complications, the prerequisites for accidents to occur during operation of existing disposal sites, and scenarios of hypothetical accidents show that the consequences of complications and accidents during the disposal of industrial and liquid radioactive wastes cannot be catastrophic. The impacts are limited by sanitary protection zones, by the fact that complications (problems) and accidents develop very slowly, and because the conditions for their occurrence can be detected by monitoring the geologic medium and the engineered systems. The greatest potential danger attaches to the engineered systems: piping, wells, and pumps. The most frequent causes of complications and accidents are poor workmanship (concealed defects), human error, and failure to take timely action to prevent the development of complications. Experience in the operation of deep injection disposal sites has confirmed that the geologic formations themselves are reliable and effective isolators that form a dependable final site for radioactive wastes.

The greatest potential danger attaches to the engineered systems: piping, wells, and pumps. The geological formations themselves are reliable and effective isolators that form a dependable final site for radioactive wastes.

6 Shutdown of Liquid Radioactive Waste Disposal Sites

Shutdown is an obligatory stage in operating deep injection disposal sites for radioactive wastes, whether liquid or solid. Shutdown and postshutdown operations must prevent detrimental impacts of radioactive wastes on human beings and on the sphere of human habitation after the wastes have been emplaced.

The main objective in shutting down radioactive waste disposal systems is to create conditions for habitation and economic activity that are as nearly identical as possible to normal conditions prevailing in areas where there has been no waste disposal. Succeeding generations must not experience any kind of inconvenience, let alone direct effects, stemming from previously buried wastes.

To achieve this goal it is necessary to first comply with the requirements for localization, compatibility, and monitoring (see Section 3.1). Wastes must be localized within the defined boundaries of the subsurface exclusion zone, the region inside which all uses of the subsurface space are restricted. The drilling of wells and the driving of mine workings that would expose the zones of waste localization are forbidden.

Surface structures at waste disposal sites, including pipes, well pavilions, and pumping and other processing buildings, must be dismantled or converted to other uses. Wells that could serve as routes by which the zone of waste localization can communicate with the surface must be stabilized so that waste components cannot move beyond the boundaries of the subsurface exclusion zone.

Similar approaches are used to shut down the facilities of economic activity having to do with the use of subsurface resources, including mineral extraction.[65] The requirement is not that subsurface conditions be restored to original conditions—this is not practically feasible. However, the surface must be left in a condition suitable for a variety of activities.

In analyzing the problems of shutting down sites where subsurface resources have been used for various purposes, including the disposal of radioactive wastes, we must consider the probable impossibility of totally preventing the consequences of prior subsurface activities from affecting the living conditions of later generations. For example, there will be limitations on the use of subsurface resources, and these may influence the economic

The main objective in shutting down radioactive waste disposal systems is to create conditions for habitation and economic activity that are as nearly identical as possible to normal conditions...

development of the area. The subjective factor will also be significant: The knowledge that wastes are present beneath the ground and awareness of their hazard potential may breed apprehensiveness that will lower the "quality of life."

In this connection, compensatory measures must be effected at waste disposal sites in their postshutdown phase. The aim of these practices must be to regulate economic activity and inform the populace.

To ensure the public is adequately informed, the geologic medium is monitored after the shutdown of a radioactive waste disposal facility. The condition of the wastes, the subsurface waters, the rocks, and the surface will be monitored. The acquired data will yield objective information for presentation to local government agencies and the populace. This information can be used to forestall the development of social tension and other negative consequences brought about by the hazard potential of the wastes.

Monitoring of disposal sites may form part of the federal or regional environmental monitoring program, which must cover the entire area of a country or region and must take in all environmental objects that may be subject to any impact.

Planning for the postshutdown phase of liquid radioactive waste disposal sites includes consideration of three groups of structures and natural objects involved in the disposal system. These groups differ in the nature of the postshutdown actions required:

1. the wastes themselves and the reservoir formations that contain them, together with the relatively impermeable and buffer horizons overlying and underlying these

2. the wells drilled at the disposal site

3. the surface equipment.

Under the fundamental principles that apply to the deep injection disposal of liquid radioactive wastes, no special postshutdown actions are required to be carried out in the reservoir formation. The selection of a geologic formation meeting certain requirements and the identification of disposal schemes and operating modes will ensure waste localization within the subsurface exclusion zone for a specified period of time. During preparation for shutdown, however, it is also necessary to demonstrate that requirements were met; this is done by citing monitoring and research data and site operating records that bear on waste disposal processes and the behavior of wastes in the subsurface. Operating records must also be used to correct the previously defined boundaries of the subsurface exclusion zone, the volume of which may be decreased for the postshutdown period, with a corresponding relaxation of restrictions on the use of subsurface resources.

When site operation and waste injection come to a close, subsurface processes will change. At the MCC and SCC, downhole pressures and the levels of formation liquids are close to natural conditions; that is, they are established below the Earth's surface. At the NIIAR site, because the density of the disposed wastes is lower than the density of the formation brines, special measures are needed to reduce the pressure head in the injection wells during the final stage of site operations. Saline solutions of increased density are injected; subsurface waters from distant observation wells or the buffer horizon can be used for this purpose.

The wellhead pressure and the formation pressure can be reduced by eliminating or substantially decreasing the conditions for vertical leaks between the reservoir formation and shallower horizons near the wells. The rate of movement of waste components in the pore space of the reservoir is diminished; this promotes the transfer of nuclides and other waste components into dead-end pores and the mineral skeleton of the reservoir rocks, and thus aids the retention of the nuclides by the rocks. The temperature of the reservoir formation begins to fall off 1 to 1.5 years after the cessation of injection of the weakly acidic process wastes arbitrarily classified as high-level wastes.

The conditions for waste localization in the reservoir formation improve after site operations are ended. Even so, the planned shutdown actions must include measures to limit waste dispersal in the most dangerous directions. These requirements grow out of several circumstances, including 1) doubts—arising among specialists and community leaders who are unfamiliar with the behavior of radioactive wastes under subsurface conditions—about the needed isolating qualities of geologic formations, and 2) fears that the rate of recent geologic processes associated with the natural development of the Earth's crust may increase. Such apprehensions are based on the opinion, stated by some scientists in recent years, that tectonic activity is intensifying and can affect the conditions of waste isolation. This effect will obviously proceed very slowly over extended periods of time, millennia according to estimates.

In this connection, work has begun on creating a method for setting up antimigration curtains or barriers; these can also be considered under the class of accident responses. The most effective response is to create physical-chemical (or geochemical, sorptional) curtains or barriers enabling the rocks to delay the migration of contaminants without appreciably upsetting the natural flow of subsurface waters or waste filtrate. In a reservoir formation, a constructed barrier used to block the flow or drastically change its direction will not be effective, because of the large contaminant volumes, the width of the flow, and the relatively great depth of the reservoir formation.

Well condition largely governs the reliability of subsurface waste localization, both during and after site operation.

One way to create a physicochemical antimigration barrier is to activate the natural retentive power of the geologic medium by treating the reservoir rocks with solutions and gas mixtures injected through existing or specially constructed wells. Reagents can also be introduced directly into the waste contour, causing precipitation in the pore space so as to trap and coprecipitate the nuclides. A number of reagents and gas mixtures have been tested in the laboratory with positive results.

Wells drilled at disposal sites are the most critical structures. Well condition largely governs the reliability of subsurface waste localization, both during and after site operation. After shutdown, wells must be left in a condition that ensures reliable isolation of all horizons that were exposed. This vital task is extremely difficult because wells are engineered systems whose structural elements are made of materials that do not occur naturally in geologic formations. Such materials would be unstable over long (geologic) spans of time. This statement applies most of all to casing strings, which are made of various alloys.

To restore the natural conditions at a well site and ensure the reliable isolation of wastes, it is necessary to use plugging-back materials whose composition and properties match natural formations as closely as possible. Such materials include a range of cements and concretes, bentonite, and zeolites. Plugging-back materials based on portland (hydraulic) cements are composed chiefly of substances similar to minerals. Portland cements are widely used in well construction.

Under the leadership of Professor V. I. Sidorov, the Moscow Engineering and Construction Institute studied the corrosion resistance of cement made from portland cement under the conditions prevailing in sand-clay horizons containing low-mineral-content hydrocarbonate waters. These conditions are typical of the relatively impermeable confining beds, buffer horizons, and shallower horizons found at the SCC and MCC disposal sites. The isolation of these horizons in intervals above the reservoir formation must be sufficiently reliable.

Based on this study, the depth to which hardened cement will fracture under the action of subsurface hydrocarbonate waters was predicted: over 100 years, the depth would be 7.7×10^3 cm. Corrosion processes are found to be significantly slowed over time. Corrosion processes in subsurface waters having the stated composition are slower than in distilled water. Microporometric studies have shown that the cement contains virtually no pores of dangerous size after testing. Diffusion processes are hindered as a result.

A mineralogical phase analysis showed that the grains of cement clinker continue to be hydrated over time and that calcium hydrosilicates and hydroaluminates continue to crystallize. As a result of these processes, cement specimens grow in compressive strength under the action of hydrocarbonate media for two reasons: 1) The porosity of the hardened cement decreases as the pores become plugged with calcium carbonate. 2) Calcium hydrosulfate aluminates are converted to calcium hydrocarbonate aluminates by substitution in an unchanging microstructure, which prevents the development of internal stresses leading to strains in the hardened cement. The stability of cement under the stated conditions is aided by the constant temperature and moisture in the deep horizons lying above the reservoir formation and in contact with the cement; these formations survive for centuries or millennia.

The Institute of Physical Chemistry studied the stability of various plugging-back materials with respect to the radiation, chemical, and thermal actions of the wastes, that is, for reservoir intervals containing waste components. Under such conditions, it is desirable to use organic polycondensation composites: thermosetting adhesives, epoxy and polyether adhesives based on epoxy resins, furfural-cationic resins, and furan-epoxy resins.

The use of "altines," which are derived from products of shale chemistry, has attracted some interest. Altines are components of polymer concretes, which combine the qualities of organic and inorganic binders. Among new and promising materials is "polimin," synthesized at the Institute of Mechanics of Moscow University. Polimin is a composite of bentonitic clay and a high-molecular-weight polymer. Research on plugging-back materials continues.

The actual method used to shut down wells is quite important for the postshutdown phase. The technology must guarantee that plugging-back material is delivered to the most critical parts of the casing-borehole annulus and the interior of the borehole.

The shutdown of deep wells, including oil and gas wells and those used for other purposes, is a widespread operation usually called "liquidation." Gosgortekhnadzor has issued standards, requirements, and methods for shutting down various wells, including special-purpose wells used for waste disposal.[62] But the great hazard potential of radioactive wastes means that special technologies have to be devised.

The shutdown of wells used for injection disposal of liquid radioactive wastes must be preceded by an inspection to determine, for example, the degree of plugging (colmatage) of the reservoir formation, the condition of hardened cement in the borehole-casing annulus, and the integrity of the string. A broad palette of geophysical and hydrogeologic methods is used: temperature measurement, acoustic cementometry, radioactive logging, location of subsurface water levels, determination of changes in geophysical characteristics in the reservoir formation and shallower horizons due to the injection of wastes and tracer solutions, and sampling and analysis of formation liquids. In complicated cases, a special monitoring well may be drilled during site shutdown. The results of this examination make it possible to justify an appropriate shutdown technology.

One of the first steps in shutting down a well is to inject plugging-back solutions into the waste-containing reservoir formation. These solutions form a practically impermeable zone in the reservoir a few meters away from the well. As a result, the wastes or the waste filtrate will not later contact the well and plugging-back materials. The technology for plugging back a well is governed by the well design and condition. At the SCC, scheduled preventive maintenance was accompanied by the experimental shutdown of some wells at the disposal site.

The wells to be shut down can be divided into two groups: 1) wells in which the cement in the borehole-casing annulus is in good condition and the reservoir formation is well isolated from shallower horizons, and 2) wells that need additional isolation of the borehole-casing annulus in the course of shutdown. No special difficulties were encountered in shutting down wells of the first group. In the second group, the special technology illustrated in Figure 42 had to be used.

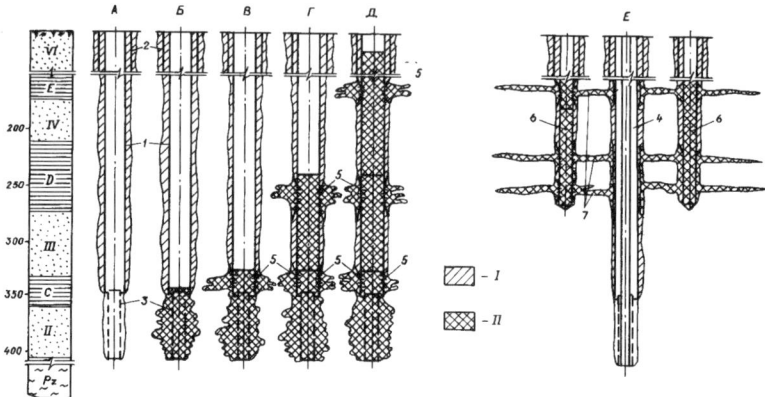

Figure 42. Shutdown of Wells at Sites Used for Deep Injection Disposal of Liquid Radioactive Wastes. (a) well to be shut down, (b) cementing of the screen zone of the well, (c) perforation of the lower section of casing and cementing in hydrofacturing mode, (d) similarly for the middle section of casing, (e) similarly for the upper section, (f) use of injection wells in shutdown. 1 - operational casing string, 2 - surface casing, 3 - screen, 4 - lift or instrumented string, 5 - perforation intervals, 6 - injection wells, 7 - cracks due to hydrofracturing of formation. I) cement emplaced when well was constructed; II) cement emplaced at shutdown time.

In wells giving access to downhole instruments, the casing string was perforated by a shaped-charge method in previously identified intervals of relatively impermeable rocks. The plugging-back mix was then injected to effect hydrofracturing of the formation. The cement thus penetrated into the borehole-casing annulus and into the relatively impermeable rocks, forming horizontal circular barriers in them. The distribution of the hardened cement was checked on the basis of the volume of cement injected and by a set of geophysical and hydrodynamic methods. A monolithic cement body including the casing strings was formed around the well. The interior of the well was also filled with cement. The horizontal cement barriers provide further isolation of the reservoir formation from shallower horizons.

For wells where access to the interior is hindered, for example because of severe contamination, a plugging-back method involving injection wells was tested. The injection wells were created 10 to 15 m from the well to be shut down; they penetrated the shallower horizons. Cement was injected under hydrofracturing conditions in intervals of relatively impermeable rocks. According to monitoring observations, the circular barriers that formed extended to the borehole of the well being shut down and filled parts of its borehole-casing annulus.

The final step in shutting down a disposal site for liquid radioactive wastes is to rehabilitate the land. There will be no appreciable difficulties here, because there is very little or no contamination of the soil associated with site operation.

It must be emphasized that the method of shutting down liquid radioactive waste disposal sites is still in the research and development stage. Before shutdown plans can be completed, interdisciplinary studies and experimental and pilot studies have to be carried out on a broad scale.

The final step in shutting down a disposal site for liquid radioactive wastes is to rehabilitate the land. There will be no appreciable difficulties here, because there is very little or no contamination of the soil associated with site operation.

7 Conclusion

The main outcome of deep injection disposal of liquid radioactive wastes can be stated as follows: the radioactive wastes at three large nuclear enterprises have been prevented from affecting nearby human beings, flora, fauna, surface waters, and shallow groundwaters. As a result hundreds, perhaps thousands, of people have avoided exposure to additional doses of radiation during the 1960s, 70s, and 80s. Consider the following important points:

- Wastes containing hundreds of millions of curies of radioactive substances have been localized in the geologic medium within sanitary protection zones and subsurface exclusion zones. Most of these radioactive nuclides are now in solid phase in the rocks, in the form of sorbate and poorly soluble compounds formed by physical and chemical processes.

- The structure and properties of the geologic medium at disposal sites and the ability of the subsurface region to hold and localize wastes have been well studied in preliminary geologic exploration and laboratory research. They have also been directly determined in the course of waste disposal. As a result, the ultimate behavior of the wastes can be predicted with confidence.

- The condition of the wastes and the geologic formations containing them are carefully monitored. These observations and measurements demonstrate that the disposal sites are safe. Another important reason for monitoring the geologic medium is to take measures for further isolation of the wastes if this becomes necessary.

- Possible complications and accidents that might occur during disposal operations would not have the catastrophic consequences often linked with large-scale impacts on ecosystems, the atmosphere, and surface waters.

- The exclusion of waste-containing portions of the geologic medium from other possible uses does not result in significant losses of natural resources, since the regions in question do not have any especially valuable qualities and do not contain scarce minerals or large reserves of subsurface waters.

- Once waste disposal has been terminated, the disposal sites can effectively be shut down.

Experience in deep injection disposal of liquid radioactive wastes, together with observations of waste dispersal and related processes, is applicable to solidified waste burial in relatively impermeable geologic formations, to preventing subsurface water pollution and forecasting the consequences of such pollution, and developing mineral deposits.

The development of technologies for conditioning and solidification of liquid radioactive wastes, accompanied by concentration and reduction of volume, will make it possible to abolish the underground disposal of radioactive wastes in liquid form. Both the surface storage and underground burial of solidified wastes can be fairly safe. The possibility of underground disposal should not, however, be ruled out when conditions are favorable for certain types of liquid radioactive wastes, such as those containing tritium and other short-lived nuclides.

The deep injection disposal of nonradioactive industrial wastes is an effective method of preventing the pollution of surface waters and streams, especially in regions with complex ecological settings and for processes whose wastes cannot be rendered innocuous by other methods. This approach to waste management is widely used in the U.S. and is beginning to be practiced in Russia. Experience with the deep injection disposal of liquid radioactive wastes, the topic of this book, will certainly be of interest in dealing with this difficult problem.

8 References

1. *Trudy vtoroi Mezhdunarodnoi konferentsii po mirnomu ispol'zovaniyu atomnoi energii, Zheneva, 1959; Izbrannye doklady inostrannykh uchenykh, Radiobiologiya i radiatsionnaya meditsina* [Proceedings of the second international conference on peaceful uses of atomic energy, Geneva, 1959; Selected papers by foreign participants, radiobiology and health physics.] Moscow, 1959.

2. A. Cleboch, Jr., and E. H. Baltz, "Progress in the United States of America toward deep-well disposal of liquid and gaseous radioactive wastes," *Proc. IAEA Symp. Disposal of Radioactive Wastes into the Ground*, pp. 591-604, IAEA, Vienna, 1967.

3. "Disposal of radioactive grouts into hydraulically fractured shale," Technical Report Series No. 232, IAEA, Vienna, 1983.

4. I. B. Robertson, *Digital modeling of radioactive and caste transport in the Snake River Plain Aquifer at the National Reactor Testing Station, Idaho*, USGS Open File Report 74-1089, 1974. Available through the National Technical Information Services, IDO-22054. Call (703) 487-4650.

5. "Sanitarnye pravila i tekhnicheskie usloviya ekspluatatsii i konservacii glubokich hranilich zhidkikh radioaktivnykh otkhodov predprijatii jaderno-toplivnogo cikla," SP and TU EKH-93 [Sanitary rules and engineering specifications for operation and shutdown of deep injection sites for liquid radioactive waste enterprises of nuclear-fuel cycle.] VNIPI-promtechnologii, Moscow, 1995.

6. "Normy radiatsionnoi bezopasnosti NRB-76/87," [Radiation safety standards NRB-76/87], Energoatomisdat, Moscow, 1988.

7. "Waste Isolation Safety Assessment Program." Scenario analysis methods for use in assessing the safety of the geologic isolation of nuclear waste, November 1978. Prepared for the Office of Nuclear Waste Isolation, U.S. Department of Energy under Contract Et-76-C-06-1830, Pacific Northwest National Laboratory, Richland, Washington.

8. G. B. Poluektova, Yu. V. Smirnova, and I. D. Sokolova, "Obrabotka i udalenie radioaktivnykh otkhodov predpriyatii atomnoi promyshlennosti zarubezhnykh stran," [Conditioning and disposal of radioactive wastes from nuclear industry enterprises in foreign countries], *Obzor TsNIIAtominform*, Moscow, 1990.

9. "Standardization of radioactive waste categories," Technical Report Series No. 101, IAEA, Vienna, 1970.

10. "Sanitarnye pravila obrashcheniya s radioaktivnymi otkhodami (SPORO085)," [Sanitary rules for handling radioactive wastes (SPORO085)], SanPiN 42-129-11-3938-85, Minzdrav USSR, Moscow, 1986.

11. A. S. Nikiforov, V. V. Kulichenio, and M. I. Zhikharev, *Obezvrezhivanie zhidkikh radioaktivnykh otkhodov* [Processing of liquid radioactive wastes], EAI, Moscow, 1985.

12. "Site investigations for repositories for solid radioactive wastes in deep continental geological formations," Technical Report Series No. 215, IAEA, Vienna, 1982.

13. K. B. Poluektova, J. V. Smirnov, and J. D. Cokolova. "Obrabotka ii udalenie radioaktivnykh otchodov predpriyaty atomnoy promyshlennosti Zarubeghnykh stran," Vol. II, p. 304, Table 6.2, Obsor, TsNiiAtominform, Moscow, 1990.

14. V. I. Zemlyanukhin, et al., *Radiokhimicheskaya pererabotka yadernogo topliva* AES [Radiochemical processing of nuclear fuel from nuclear power plants], Energoatomizdat, Moscow, 1983.

15. V. V. Gromov, et al., "Khimicheskaya tekhnologiya obluchennogo yadernogo goryuchego" [Chemical tech-nology of irradiated nuclear fuel], *Atomizdat*, Moscow, 1971.

16. "Gosudarstvennaya programma Rossiiskoi Federatsii po radiatsionnoi reabilitatsii Ural'skogo regiona i merakh po okazaniyu pomoshchi postradavshemu naseleniyu na period do 1995 g," [The state program of the Russian Federation for radiation rehabilitation of the Ural region and measures for extending assistance to impacted population in the period up to 1995], Postanovlenie Verkhovnogo Soveta R.F., po 5148 ot 10.06, Moscow, 1993.

17. S. I. Beard and W. L. Codfrey, "Waste disposal into the ground at Hanford," *Proceedings IAEA Symposium on Disposal of Radioac-tive Wastes into the Ground*, pp. 123-132, IAEA, Vienna, 1967.

18. H. Brücher, "Contributions to the risk evaluation of a high-level waste solidification plant," *Nuclear Technology*, vol. 39, 1978.

19. "Rekomendatsii Mezhdunarodnoi komissii po radiologicheskoi zashchite," [Recommendations of the International Commission on Radiological Protection], Publication 2, GAI, Moscow, 1961.

20. "Doklad Nauchnogo komiteta OON po deistviyu atomnoi radiatsii" [Report of the UN Scientific Committee on effects of atomic radiation], A/5216, New York, April 1962.

21. "Safety principles and technical criteria for the underground disposal of high level radioactive wastes," Safety Ser. No. 99, IAEA Safety Standards, IAEA, Vienna, 1989.

22. N. I. Plotnikov, *Podzemnye vody-nashe bogatstvo* [Subsurface waters, our riches], Nedra, Moscow, 1990.

23. *Spravochnoe rukovodstvo gidrogeologa* [Hydrogeologist's handbook], edited by V. M. Maksimov, Vol. 1, Nedra, Leningrad, 1979.

24. V. M. Gol'dberg and N. P. Skvortsov, *Pronitsaemost' i fil'tratsiya v glinakh* [Permeability and flow in clays], Nedra, Moscow, 1986.

25. V. P. Gavrilov, *Proiskhozhdenie nefti* [The genesis of petroleum], Nedra, Moscow, 1986.

26. *Neftyanye i gazovye mestorozhdeniya SSSR, Spravochnik* [Oil and gas fields of the USSR, a handbook], Nedra, Moscow, 1987.

27. *Sovremennye metody izucheniya i prognoza pokryshek nefti i gaza* [Current methods of studying and predicting oil and gas cap rocks], Minsk, 1981.

28. G. E. Prozorovich, *Pokryshki zalezhei nefti i gaza* [Cap rocks of oil and gas pools], Nedra, Moscow, 1972.

29. V. S. Kovalev and V. I. Zhitomirskii, *Prognoz razrabotki neftyanykh mestorozhdenii i effektivnost' sistem zavodneniya* [Predicting the development of oil fields and the effectiveness of water flooding systems], Nedra, Moscow, 1976.

30. I. E. Evseeva, A. I. Perel'man, and K. E. Ivanov, *Geokhimiya urana v zone gipergeneza* [Geochemistry of uranium in the zone of hypergenesis], Gosatomizdat, Moscow, Minatomiadat, 1974.

31. *Sovremennye dvizheniya zemnoi kory* [Recent movements of the Earth's crust], Nauka, Moscow, 1987.

32. *Poyasnitel'naya zapiska k karte neotektonicheskogo raionirovaniya Nechernozemnoi zony RSFSR, M. 1:1500000* [Explanatory notes to the 1:1,500,000 map of neotectonic zonation of the non-Chernozem zone of the RSFSR], Mingeo RSFSR and Moscow University, Moscow, 1983.

33. A. A. Nikonov, *Zemletryaseniya, Proshloe, sovremennost', prognos* [Earthquakes· the past, the present, prediction], Znanie, Moscow, 1984.

34. Kh. Shakh, *Zybkaya tverd', Chto takoe zemletryasenie i kak k nemu podgotovit'sya* [Unfirm ground: What is an earthquake and how to prepare for it], Mir, Moscow, 1988.

35. D. A. Eibi, *Zemletryaseniya* [Earthquakes], Nedra, Moscow, 1982. [Russian translation of G.A. Eiby, *Earthquakes.*]

36. *Gidrogeologicheskie issledovaniya dlya zakhoroneniya promyshlennykh stochnykh vod v glubokie vodonosnye gorizonty* [Hydrogeologic studies for the disposal of industrial waste-waters in deep aquifers], (Metodicheskie ukasania), edited by K. J. Antonenko and E. J. Chapovskogo, Nedra, Moscow, 1976.

37. *Gidrogeologicheskie issledovaniya dlya obosnovaniya podzemnogo zakhoroneniya promyshlennykh stokov* [Hydrogeologic studies for the justification of the underground disposal of industrial wastes], edited by V. N. Grabovnikov, Nedra, Moscow, 1993.

38. V. I. Spitsyn and V. D. Balukova, "Investigation of sorption and migration of radioisotopes in soils and rocks of different compositions," *Proceedings IAEA Symposium on Disposal of Radioactive Wastes into the Ground*, pp. 169-177, IAEA, Vienna, 1967.

39. Shokei Kata and Hajimu Jabuta, "Distribution coefficients used for safety assessment for shallow land radioactive waste burial," *Nihon Geushiryoku Gakkaaishi*, isbb, 28(4).

40. Yu. V. Kuznetsov and V. N. Shchebetkovskii, *Osnovy ochistki vody ot radioaktivnykh zagryaznenii* [Principles of removing radioactive contaminants from water], Atomizdat, Moscow, 1974.

41. Yu. I. Tarasevich and F. D. Ovcharenko, *Adsorbtsiya na glinistykh mineralakh* [Adsorption on clay minerals], Naukova dumka, Kiev, 1975.

42. Yu. A. Ol'khovik and E. V. Sobotovich, "Perenos radioaktivnykh nuklidov v protsesse fil'tratsii iz vodoema," [Transport of radioactive nuclides in the process of flow from a body of water], Zh. Vodnye resursy/6, 1990.

43. "Assessment of migration pathways," Tech. Symp. Stand. for High-Level Radioactive Waste Management, EPA Contract No. 68-01-4470, U.S. Environmental Protection Agency, 1977.

44. V. N. Shchelkachev and B. B. Lapuk, *Podzemnaya gidravlika* [Subsurface hydraulics], Gostoptekhizdat, Moscow, 1949.

45. V. N. Shchelkachev, *Razrabotka neftevodonosnykh plastov pri uprugom rezhime* [Development of oil- and water-bearing formations in elastic regime], Gostoptekhizdat, Moscow, 1959.

46. P. Ya. Polubarinova-Kochina, *Teoriya dvizheniya gruntovykh vod* [Theory of groundwater movement], Gostekhteorizdat, Moscow, 1952.

47. F. M. Bochever and A. E. Oradovskaya, *Gidrogeologicheskoe obosnovanie zashchity podzemnykh vod i vodozaborov ot zagryaznenii* [Hydrogeologic justification of protection of subsurface waters and water intakes from pollution], Nedra, Moscow, 1976.

48. V. M. Gol'dberg, *Gidrogeologicheskie prognozy dvizheniya zagryaznennykh podzemnykh vod* [Hydrogeologic predictions of the movement of polluted subsurface waters], Nedra, Moscow, 1973.

49. V. A. Mironenko, *Dinamika podzemnykh vod* [Dynamics of subsurface waters], Nedra, Moscow, 1983.

50. L. Lukner and V. M. Shestakov, *Modelirovanie migratsii podzemnykh vod* [Modeling of migration of subsurface waters], Nedra, Moscow, 1986.

51. N. N. Verigin, "O skladirovanii zhidkikh produktov i otkhodov promyshlennosti v poristo-treshchinnykh gornykh porodakh," [Deposition of liquid industrial products and wastes in porous-fissured rocks], *Izvestiya vysshikh uchebnykh zavedenii, Geologiya i razvedka*, No. 10, 1968.

52. P. P. Kostin, "Nekotorye osobennosti gidrogeologicheskikh protsessov pri podzemnom zakhoronenii promstokov," [Some features of hydrogeologic processes in underground disposal of industrial wastes], *Izv. VUZov, Geologiya i razvedka*, No. 11, 1989.

53. A. S. Belitskii, *Okhrana okruzhayushchei sredy pri podzemnom zakhoronenii promstokov* [Environmental protection in underground disposal of industrial wastes], Nedra, Moscow, 1976.

54. L. Lukner and V. M. Shestakov, *Modelirovanie geofil'tratsii* [Modeling of geofiltration], Nedra, Moscow, 1976.

55. V. A. Mironenko and V. G. Rumynin, *Opytno-migratsionnye raboty v vodonosnykh plastakh* [Experimental migration studies in aquifers], Nedra, Moscow, 1986.

56. V. M. Dobrynin, *Deformatsii i izmeneniya fizicheskikh svoistv kollektorov nefti i gaza* [Deformations and changes in the physical properties of oil and gas reservoirs], Nedra, Moscow, 1970.

57. O. L. Kedrovskii, et al., "Printsipy otsenki nadezhnosti podzemnogo zakhoroneniya radioaktivnykh zhidkikh otkhodov v glubokie geologicheskie formatsii i puti ee povysheniya" [Principles of assessing the reliability of underground disposal of liquid radioactive wastes in deep geologic formations and methods for improving reliability], in *Underground Disposal of Radioactive Wastes*, IAEA, Vienna, 1980.

58. "Nadezhnost' v tekhnike, Terminy i opredeleniya," [Reliability in engineering, terms, and definitions], GOST 27.004-85 Gosudarstvennji Komitet USSR po standartam, Moscow, 1995.

59. "Okhrana okruzhayushchei sredy, Postateinyi kommentarii k zakonu Rossii," [Environmental protection, section-by-section commentary on a Russian law], *Zakonodatel'stvo i ekonomika,* Nos. 16 and 17 (38), 1992.

60. *Gidrogeologicheskoe prognozirovanie* [Hydrogeologic prediction], Mir, Moscow, 1988.

61. V. I. Spitsyn, et al., "Osnovnye predposylki i praktika ispol'zovaniya glubokikh vodonosnykh gorizontov dlya zakhoroneniya zhidkikh radioaktivnykh otkhodov," [Basic prerequisites and practice of the use of deep aquifers for the disposal of liquid radioactive wastes], *Zh. Atomnaya energiya,* Vol. 44, No. 2, 1978.

62. O. L. Kedrovskii, et al., "Glubinnoe zakhoronenie zhidkikh radioaktivnykh otkhodov v poristye geologicheskie formatsii," [Underground disposal of liquid radioactive wastes in porous geologic formations], *Atomnaya energiya,* Vol. 70, No. 5, 1991.

63. I. Miroshnikov, "Chto zhe proizoshlo v Dimitrovgrade?" [What has happened in Dimitrovgrad?], *Ul'yanovskaya pravda,* October 6, 1990.

64. V. N. Strakhov, et al., "A yadernykh zemletryasenii ne bylo" [There were no nuclear earthquakes], *Ul'yanovskaya pravda,* December 1, 1990.

65. "Polozhenie o poryadke likvidatsii neftyanykh, gazovykh i drugikh skvazhin i spisaniya zatrat na ikh sooruzhenie," [Decree on the procedure for shutting down oil, gas and other wells and amortization of well construction expenses], ratified by Gosgortekhnadzor, December 27, 1989.

Index

Note: Pages where the indexed information is in the figure are denoted by F. Pages where the indexed information is in the table are denoted by T.

1 MONTH OF
FREE
READING

at

www.ForgottenBooks.com

By purchasing this book you are eligible for one month membership to ForgottenBooks.com, giving you unlimited access to our entire collection of over 1,000,000 titles via our web site and mobile apps.

To claim your free month visit:

www.forgottenbooks.com/free45692

ISBN 978-1-5280-5420-1
PIBN 10045692

This book is a reproduction of an important historical work. Forgotten Books uses state-of-the-art technology to digitally reconstruct the work, preserving the original format whilst repairing imperfections present in the aged copy. In rare cases, an imperfection in the original, such as a blemish or missing page, may be replicated in our edition. We do, however, repair the vast majority of imperfections successfully; any imperfections that remain are intentionally left to preserve the state of such historical works.

THEY WHO UNDERSTAND

BY

LILIAN WHITING

"Be constant, O happy soul, be constant and of good courage! For thou wilt be protected, enriched, and enlightened by the greatest good; and if thou dost not turn away, but perseverest constantly, know that thou offerest to God the most acceptable sacrifice." — MIGUEL MOLINOS.

BOSTON

LITTLE, BROWN, AND COMPANY

1919

Norwood Press

Set up and electrotyped by J. S. Cushing Co., Norwood, Mass., U.S.A.

Presswork by S. J. Parkhill & Co., Boston, Mass., U.S.A.

To

THE BELOVED AND PRECIOUS MEMORY

OF

THE FLOWER OF AMERICAN YOUTH

WHOSE HEROISM EXALTS AND CONSECRATES

THE NEW FREEDOM

THAT WILL INVEST A REMADE WORLD

THUS TRANSFIGURED BY

THEIR HOLY SACRIFICE

—LILIAN WHITING

" The gift of God is eternal life through
Jesus Christ, our Lord "

CONTENTS

" There shall never be one lost good! What was, shall live
 as before ;
The evil is null, is naught, is silence implying sound ;
What was good shall be good, with, for evil, so much good
 more ;
On the earth the broken arcs; in the heaven a perfect
 round."

     —BROWNING in "Abt Vogler."

THEY WHO UNDERSTAND

I

THE GATES OF NEW LIFE

"... a Hand like this hand
Shall throw open the gates of new life to thee!
See the Christ stand!"—BROWNING in "Saul."

A GREAT spiritual awakening is over the world. "Where Christ brings His cross He brings His presence," and never was the intuitive turning of all humanity to God, in the face of sorrow, more evident than at the present time. We read a new meaning into the wonderful words, "God is our refuge and our strength; a very present help in time of trouble." The words are a foundation of actual life; not merely nor even mostly consolation, in the ordinary sense, but a basis of the deepest reality on which to stand. We endure — as seeing the invisible. It is the world we do not see in which we live; it is the forces of the unseen which sustain all purpose. Nor is it only in hours of sadness and bereavement that we would turn to God; our own poet of the spiritual life, the gentle and beloved Longfellow, has given true expression to an universal feeling in the lines:

"Ah, when the infinite burden of life descendeth
    upon us,
Crushes to earth our hope, and, under the earth,
    in the graveyard,
Then it is good to pray unto God! for His sorrowing children
Turns He ne'er from His door, but He heals and
    helps and consoles them.
Yet it is better to pray when all things are prosperous with us,
Pray in fortunate days, for life's most beautiful
    Fortune
Kneels before the Eternal's throne; and with
    hands interfolded,
Praises thankful and movéd the only Giver of
    blessings."

A very present help in time of trouble, a help equally needed in time of joy, — in every supreme

experience of life the soul turns intuitively and
instinctively to the divine aid. The nature of
this aid is constantly being more clearly revealed
to us. It is also true that in the deepening
spirituality of life man is more and more depending on this aid. Our religious faith is becoming
to us the most absolutely practical reliance. This
deeper assurance springs largely from our increasing
comprehension of the nature of life; of the origin,
the development, the conditions of progress, and
the final destiny of the spiritual man which is
the individual himself. To speak of the destiny
of the soul as if it were something apart from the
man, is misleading. Shall we not realize the
simple and fundamental truth that we are, here
and now, spiritual beings, dwelling in a spiritual
world; that it is the spiritual and not the physical
world to which we belong; that we are temporarily clothed with a physical body as the instrument in correspondence with the physical
environment in which we sojourn for a season?
Yet, all the while, even during this sojourn, we
are still the inhabitants of the spiritual world;
a world of "discrete degrees", as Swedenborg

points out, in which the ethereal is the next succeeding environment to the physical; after which we pass on to still finer and finer degrees of environment, even from glory to glory, as the apostle phrases it. Now, as we are here and in the immediate present an inhabitant of both the physical and ethereal realms; tethered to the former by the physical mechanism; related to the latter by virtue of the ethereal body in which we find ourselves when we withdraw from the physical body (as one would withdraw his hand from a glove), does it not seem luminously clear that those of our beloved who have thus withdrawn by the process we name death are still in close relations to us? Never was there a time in human history when the question was so vital as now, when thousands of homes are desolated by the vanishing of son, brother, or husband in the tragic and terrible conflict which has been raging. Unless life and all its interests and purposes extend beyond the merely visible limits, what philosophy or consolation could we find?

During the Boer War Archdeacon Wilberforce

said, in a private letter to a friend : "What do you think is the state of these great numbers of young Englishmen suddenly hurled out of life? Where are they? What are their first experiences?" When the present Archbishop of Canterbury visited the United States in 1906, preaching eloquently in many churches, he asked, in one discourse, "The life beyond, — what is it? What is its relation to the life about us?" The Archbishop instanced this question as the first one that would rush to our lips if, for a single hour, we had full access to Him "who is the Source and Object of our faith."

If that question were vital in 1906, what is it in this year of 1919, when it voices the thought that is in every heart? —We are living in great moments. Supreme sacrifice is lifting humanity to the heights hitherto undreamed. But through what suffering, what sorrow of bereavement do we strive to behold a still nobler future! Are those homes made desolate; those hearts which cry, —

"But oh for the touch of a vanished hand,
    And the sound of a voice that is still!"

are they to be left groping in chaotic darkness,
hoping, trusting, yet feeling that they do not
really know in what state are these gallant young
lives that have passed, or in what relation to life
still here?

Can we know? It is not too much to say that
it is absolutely assured that we may penetrate
to a considerable extent beyond the horizon line
that divides the unseen and the seen. For this
horizon line is not a fixed wall; it is not a definite
and immovable boundary; it is a line that recedes
as constantly before the increasing development
of spiritual perceptions as does the horizon line
of distance before the eye of the traveler. Scientific knowledge of the nature of the universe and
the increasing power to *lay hold* of spiritual truth
unite to reveal to man something of the conditions
in which those who withdraw from the physical
world find themselves. So we may question:
After all, just what has happened? One who
was on earth yesterday, so tenderly beloved and
cherished, is to-day in the next succeeding environment of our eternal and immortal life. What
does this transition signify to him and to us?

First of all, we may be confidently assured it does not signify loss and loneliness and unbroken sorrow. To a marvelous degree death gives, rather than takes away. Spirit to spirit approaches more closely than when both were limited by the physical mechanism. One is now liberated from this, and therefore more fully in command of his powers. When one comes to think of it, the physical body is a separation to a degree. How universal is the recognition of love far deeper than can be expressed in human language. How universal is the recognition of both feeling and thought that can never be fully translated into ordinary expression.

> "We are spirits, clad in veils;
>   Man by man was never seen;
>   All our deep communion fails
>   To remove the shadowy screen."

In this stanza and others in the same poem, Christopher Pearse Cranch, one of the spirits "finely touched but to fine issues", — one of that Cambridge group which included Lowell, Story, and that spirit of loveliness and love whom we

knew on earth as Charles Eliot Norton, — in these
lines Mr. Cranch felicitously embodied a profound truth. In this part of life we are veiled
to each other. We do not, at best, penetrate
very far beyond the "shadowy screen." The
tragedy of love is its possible misinterpretations.
"How often," said Mr. Longfellow, "we call a
man cold when he is only sad." As a matter of
fact there may be a beautiful interlude in this
period when one of the two closely conjoined by
ties of affection is in the ethereal and one still in
the physical world. There are thus three phases
of companionship which are fairly clear to us:
One when both are here in this part of life; the
second when one is in the ethereal and the other
here; while the third, when both are again together in the same environment in the next successive stage of life, is becoming recognizable to us.

We did not regard it as a cause for sorrow when,
in the easy and happy days that preceded that
fatal August of 1914, one held most dear left
us for a time for a journey to Europe or to the
Far East. The visible presence had temporarily
vanished, but what added richness of life was

shared! The interest and charm of the new experiences of the traveler brought their added interest and charm to the life of the one who remained at home. The analogy is unerring. The interlude of companionship between one in the unseen and one here may be, — indeed, it should be, — a period of peculiar uplifting and holy joy. One reason (perhaps the only reason) why it is not, is that the one left on earth is so plunged into grief, so submerged in sorrow, that the continual messages of thought and love cannot pass through the impenetrable gloom. Washington Irving said that sorrow for the dead was the only sorrow that we cherished; all other wounds we sought to heal, but this sorrow we regarded as one that we should not endeavor to lessen. The words reveal to how signal a degree we have advanced between the time of Mr. Irving and our own. Even when grief is unassuaged, the one in sorrow now makes brave efforts to rise above it and be cheerful for the sake of others. During the past quarter of a century the change of attitude toward death has been very apparent.

Perhaps no one who was present at the last
rites for Phillips Brooks (January 26, 1893) can
ever fail to remember that the entire spirit of the
service was that of a sacred festival.   There was
such spontaneous recognition of the immortal
qualities of the man that there was no room for
mourning.   It was felt by all that there was little
of his life that could die.   Those who have been
privileged to hold close companionship with the
noble and the lofty cannot regard their transition
from this phase of life as any finality of separation.
In all ages and in all nations the great of soul
have transcended death.   The Reverend Doctor
Ernest Stires, rector of St. Thomas's in New
York, thus speaks in a recent discourse of this
transition:

"How very stupid we are about death!   The
day that brings God's summons is our real
Commencement Day.   All our earthly life is an
education, a preparation, for a larger career.
The best that we have done here is valuable not
merely for its contribution to earthly life, but
for the training for the higher service."

Doctor Stires added:

"Hold fast to your comforting idea of God; keep your inspiring vision of life's meaning; have beautiful dreams of the joy of dear ones in the Life Eternal; and remember that all our ideas, our visions, our dreams are true only as they may be beautiful and strengthening; and that at the point of their fairest beauty they are yet short of the glorious facts, for the realities of God transcend man's highest hopes."

This interlude is one that has come into thousands of homes from which the brave and gallant youth of our country have gone forth to return no more. "This will be known as the age 'when knighthood was in flower'," Doctor Stires has also said, — the age in which the spirit of youth responded to the voice and the vision. "Life runs large" in the inspiration of a Cause when to the young man there comes that "voice without reply", and he hears, —

"'Tis man's perdition to be safe
When for the truth he ought to die."

It is a spiritual awakening to this young knighthood.

"I think I should go mad if I did not cherish faith in the justice of things, and a confident belief that death cannot end great friendships," wrote Robert Sterling, who won the Newdigate prize at Oxford for his poem, "The Burial of Socrates", and who was killed at the front on St. George's Day of 1915. This boy-poet, whose sojourn on earth had been less than twenty-two years, and Alan Seeger, who knew that he had "a rendezvous with death", and who went forward with joyful courage, are two, typical of multitudes. These young men who enter on the next phase of life are aglow with noblest enthusiasms; they are spiritually alive; they are in readiness to lay hold on progress as is the youth who enters the university filled with enthusiasm for learning rather than with indifference to his privileges. "He in whom the divine light has not awakened is virtually asleep in the spirit, and therefore cannot act upon spiritual things any more than a man asleep can act upon material things," says an Oriental writer. The conditions in which these young men pass into the unseen render them spiritually awake and alert. They compassed more than the ordinary spiritual progress of a lifetime within the brief period of their entrance into a conflict which aroused all holy enthusiasm. This fact, alone, is one of infinite significance. What new meaning has their transition? One aspect of this significance is that study and research into spiritual truth has quite established the actual fact that the higher spirituality achieved during the physical tenure of life renders the spiritual man far more free and buoyant on his entrance into the ethereal realm. The analogy may be found in that of one entering on this life with unimpeded vision rather than blindness.

After all, just what has happened? One who, so tenderly beloved and cherished, was here yesterday, sharer of our familiar conditions, is to-day in the conditions just succeeding our own; he has withdrawn from these. Yesterday he was in the physical realm. To-day he is in the ethereal realm. What does this transition signify to us, or to him?

The tragedy of the War has brought home to us these questions in a way that becomes a vital issue. Where are they, — the gallant young soldiers who offered their earthly lives with abounding heroism for the great cause of human freedom? In *The Nation*, under date of July 13, 1918, occurs this paragraph:

" Of the spiritual questions raised anew by the Great War, none is attracting more attention than that of the immortality of the soul. The enormous loss of life on the battlefield, the unfulfilled character of the lives thus abruptly ended, the hunger of those left behind for reunion with 'the loved and lost' combine to quicken and deepen the perennial interest in the problem of survival after death. Of the various phases of this interest in immortality, none is more striking than the renewal of discussion of spiritualism, psychical research, and kindred matters."

The question of the immortality of the soul! To those whose faith in immortality is as absolute as their existence, the idea of its being a "question", a debatable problem, is almost untenable. Yet that to a large proportion of humanity it still is such must be recognized. The "will to believe" does not alone create faith. That seems to be a conviction with or without which one is born. One has faith as he has his very existence; or, — he has not. Nor is it a question of ethics or morals. It is, apparently, a question of the degree of one's spiritual development, of the opening of the spiritual nature. Multitudes of people, of flawless integrity and beneficent life, do not yet find themselves with this absolute and unquestioning conviction. And as Tennyson so justly says, —

"There lives more faith in honest doubt,
    Believe me, than in half the creeds."

There is no virtue in professing a belief, a conviction, that one does not feel. Quite the contrary. Let us be honest with ourselves. Let us search, — not for argument to sustain any favorite or preconceived theory, but for truth alone. Yet as Frederic W. H. Myers has said, there is, doubtless, in each of us "an abiding psychical entity far more extensive than one knows; an individuality which can never express itself completely through any corporeal manifestation." Few are the persons who are mentally satisfied to deny the possibility of immortality, even though they declare that they perceive no evidence for it. Very few persons find themselves resting contentedly with a negative conviction. They "hope" it is true, even while, in the same breath, they may declare that they see no reason to justify this hope. In a way there seem to be three classes of attitude; that which believes unquestioningly from intuitive recognition supported by religious faith; that which has come to be convinced by evidence, — the evidence of survival by means of communications and messages from beyond; and that which is quite ready to be convinced, if the evidence seems to them sufficiently undeniable.

To no one of these attitudes can any objection be made. For they are all honest and sincere.

For more or less varying periods the matter is not, to many, the most vital issue of life. All at once through a great bereavement it becomes such. The heroic young son, brother, or husband has suddenly met death on the battlefield. Or, in some other manner, some one dearly beloved has vanished into the unseen. Then love is on the alert to penetrate the mystery.

First of all let us realize that nothing evil has happened. This change whose process we call death is simply that the spiritual man, the real being, one's self, so to speak, withdraws from the outer physical tenement, just as the hand is withdrawn from a glove. The spiritual being which is the man himself is temporarily clothed with a physical body for his use while he is in the physical world. It is this which relates him to the physical world; which enables him to come into touch with it. It is the instrument, the mechanism, that provides for the spiritual being his means of acting on and with physical forces, just as the piano, the violin, the pen, the type-writer, enable the musician to audibly embody his music, the writer to make visible expression of his thought. The physical body is that wonderful and perfectly adapted mechanism, or instrument, by which alone the spiritual being can come into relations with, and by means of which he may effectively accomplish achievements in the physical world. It is no more the man himself than the glove is the hand; or than the piano is the musician, or the pen the writer. Shall we not clearly recognize this truth, first of all? The man has withdrawn from his physical sheath. At best, it was only designed for temporary use. Somewhere within a hundred years, as a usual thing, we all withdraw from these sheaths. And then?

Then we enter on the life more abundant. But just what does that inspiring phrase signify? Is it merely a vague term whose meaning cannot be clearly grasped? Not so. The physical body, while its use is to permit the man to relate his energies to a range of objective achievements, yet limits his expression. He has far greater capabilities than can thus be expressed, as a great musician cannot adequately express his music by a piano limited to four octaves. The spiritual man then, the real individual, has far more to express than the limited mechanism of the physical body allows him to transmit through its means; therefore, when he has withdrawn from it he experiences a sense of freedom, of an exhilaration of energy, of a power undreamed of before.

The first sensation, as a rule, is that of a fairly rapturous and ecstatic delight. We know this by the vast accumulation of testimony that cannot be either doubted or denied. From the assurance of Jesus, the Christ, to the present time, its volume has been increasing. The ethereal body (which is now free from the limitations of the outer physical body) is in correspondence with the ethereal environment, the realm just succeeding that in which we now live.

What is the nature of this environment? Is it something so strange, so incomprehensible to our present conceptions, that we can form no idea at all of it? Not so. It is perfectly natural. It is in a perfect continuity of relation to our present environment. ⁻It has been called a replica of the physical world. But, instead, the physical world is a lesser and feebler replica of the ethereal. Because the latter is the more real. The ethereal is the realm of causes. The physical is the realm of effects. As life progresses it grows more real and more significant, as the life of the man or woman is more real and significant

than the life of the infant. But, holding the
analogy still further, as the infant merges into
childhood, youth, maturity, age, without any
startling change from day to day, progressing
by a system of perfect and unbroken continuity,
so, in this absolutely unbroken continuity, does
the life in the physical world merge into that
of the ethereal world. There is no definite line
of demarcation. It is the unbroken continuity of
evolutionary progression. The man who shared
our life yesterday in these familiar surroundings
shares our life to-day in his new environment.
He, in his essential self, is unchanged. But he
has entered on a larger round of possibilities and
of opportunities for his expanding powers. His
first sensation is that of an incommunicable
joy. This ecstatic sense of freedom; this intense
interest of a new and boundless range of life, —
not separated from the order of life he has just
left, but including that and beckoning on to that
which is infinitely greater, —how beautiful and
how joyous it is! With one possible exception?
Alas, it is almost always an exception, and that
is the grief of those dear to him who do not com-

prehend the blessedness and the beauty of the transition.

Now when we come to realize its true nature, should not this interlude be a joyful one on both sides? May we not think of our dear human relations as falling into three distinctive phases; the one when both are in the physical world; the second when one is in the physical, one in the ethereal; the third when both are in the ethereal? The first one of these phases has had its sweetness and its joy for us; but the second, too, has its joy and its sweetness. "Lift up your hearts." Nothing evil has happened. The companionship of spirit to spirit is unbroken. Then, the third phase, that of the reunion of both in the ethereal world, awaits. It is an event absolutely assured. There is no doubt about it. It is, at most, only a question of time. Now, why not accept the happiness, yes, even the *happiness* of this interlude? It offers its own beauty and interest. It offers great opportunities for both intellectual and spiritual experience and expansion. It has its own peculiar privileges and special joys that have not presented themselves before and will

not present themselves in just this manner again.
Shall we not make it a rich and beautiful period
rather than one of loss and gloom and sorrow?
Because in that way we may contribute so much
to the happiness of those who are so dear and who
have passed into the unseen.

It is not strange that this period has been made
one of mourning and sadness to those who have
not come to comprehend more truly the real nature
of that change we call death. It has been veiled
in mystery because we have not fully understood
the real teaching of Jesus. To some extent both
He and the apostles taught in parables and in
symbolic language, and it is only in the larger
illumination of modern interpretation that we
have quite realized the simple and sincere meaning of the gospels. "With what body do they
come?" asks Saint Paul in his epistle to the
Corinthians. The context compares the resurrection of the ethereal body with the physical
body, — the withdrawal of the ethereal from the
physical — with the sowing of grain which is
quickened and springs up from the ground.
"So, also," says Saint Paul, "is the resurrection

of the dead." Our error in the past has been that we failed to realize that his "resurrection" is but another name for the very process that we call death. It is the rising of the spiritual man from the physical encasement which he discards, as one discards outworn garments. He who dies thus rises in newness of life. That is what dying means. Now to rise in newness of life is very beautiful. It is also very joyful. And the beauty and the joy are for us whose love follows the arisen, so tenderly and unfailingly, as well as for them. Indeed, their possibilities of joy are very greatly diminished if not lost by our grief and sorrow. Now it is the one greatest comfort to feel that we may *still* do something for those dearer than our own life; and we *can* do this; we can lift up our hearts and recognize the nature of the change that has come to them and share with them the joy of it. Archdeacon Wilberforce of Westminster in an Easter sermon said: "Resurrection means continuity of individuality, utter abolition of death as a concrete reality, the exposure of death as a sham and a delusion." These are strong words from one of the most devout of churchmen; and in addition the Archdeacon suggests:

"It is mere self-deception, of course, to pretend that death is a delusion on the physical plane; it is not; . . . but, from within, the man, — the real man, rises into the new conditions. . . . The moment of death is the moment of resurrection, the essential identity the same. And remember, death is the re-uniter of loving presences."

The young hero who, in all his holy enthusiasm, flung himself into devotion to the sublime ideal for which our soldiers were fighting, and who, yesterday at the front, was separated from those who held him nearest, but who, to-day, has passed into the unseen realm, is no longer separated. Death gives us our beloved. It is the contingencies of this part of life that take them from us.

Following the wonderful illumination of the teachings of Saint Paul we read: "It is sown a natural body; it is raised a spiritual body. There is a natural body and there is a spiritual body." The two are coincident; the spiritual body (which is the real, the substantial body) is clothed by the natural for a limited period of time. One need not look beyond the familiar passages of the gospels to find authority and confirmation for the conviction of the present reality of the spiritual body (the "substantial" body, as Saint Paul well calls it), for it persists; while its outer physical case, being unsubstantial, decays and disappears on the withdrawal of the substantial one. To clearly recognize this initial fact is to gain the conditions to grasp the larger truth in direct sequence, — that, with the existence of the friend in his spiritual body (of which the physical form we knew was a replica), companionship and communion, even definite communication, are natural and even inevitable.

If I seem to dwell unduly upon this matter of the spiritual body it is because psychical research has so largely used the term "discarnate" in referring to those who have withdrawn from the life on earth. The term "discarnate spirit" may be scientific (by custom) but it is not spiritual truth. There is no such thing in the infinite universe as a "discarnate" spirit. Every spirit is clothed in a body. As life goes on and on, these bodies become finer and more subtle. But for the moment we are not considering the momentous possibilities of future eternities, but rather the immediate present after the withdrawal. For the sake of clearness may I just say that in a vast completeness of contemplation, the body that first succeeds the physical is termed the ethereal; and that successively between the conditions of the ethereal and the spiritual bodies there are differences of degree; but not to make our present survey encumbered with detail, one may simply refer to this the real body as the spiritual, which it is, indeed, in a potential degree.

The assumption that the natural grief and sorrow for the death of those tenderly cherished is a matter to be approached without comprehension and sympathy is not tenable. Into all the sweet relations of our human life this sorrow falls. It is our universal experience. But just because it is universal, a grief in common to us all, we may approach it with mutual inquiry.

A little understanding of the conditions in which we now live throws great light upon the problem of the interrelations of life in the physical and the ethereal realms. We are, here and now, spiritual beings inhabiting a spiritual world. We are only partially physical beings inhabiting a physical world. Our sojourn in this physical realm is limited. Our physical body is only a temporary convenience.

When Sir Oliver Lodge made clear to science the existence of the ether of space he thus provided a very definite condition for the environment of those who have passed through death. Sir Oliver's work was purely scientific; but one could hardly grasp the scientific truth without discerning its spiritual prototype. The great scientist finds that the ether is the most solid, the most substantial thing in the known universe, — "Perhaps the only substantial thing in the material universe," he says. Sir Oliver adds that, in comparison with the ether, "the densest matter, such as lead, or gold, is a filmy, gossamer structure; like a comet's tail, or a milky way, or like a salt in a very dilute solution." Now this substantial, etheric world is absolutely interpenetrated with our physical world. It forms

conditions coexistent. With this ethereal environment the ethereal (or spiritual) body is in
the same correspondence that the physical body
is with the physical environment. So this truth
provides a definite answer to our first question:
Where are those who were here yesterday and
have vanished to-day? Where are they? Under
what conditions are they living?

Think of the difference it is to us to simply
believe in immortality, but with no definite idea
as to what form immortality assumes; to try
to conceive a "discarnate" spirit; an "essence";
a "persistence of consciousness"; or to realize
that the man who has withdrawn from his
physical body is as definitely clothed in his ethereal
body and is living as definitely (and as naturally)
in the ethereal environment as we are in the
physical environment. What a tremendous
difference that makes to us at once. There is
something to take hold of, to understand. We
not only believe; we absolutely realize something
of the nature of the life in which he is now dwelling. Sir Oliver Lodge did not himself, in his
wonderful little book entitled "The Ether Of

Space", present its spiritual prototype. It is the purely scientific work of a great scientist. That is what makes it so tenable as a basis from which to still farther extend its significance. For if this ethereal world is so substantial one recognizes that it provides and explains the environment for the next succeeding phase of life.

That communication exists between those in the seen and those in the unseen worlds is a truth as definitely and unmistakably proven as is the reality of messages by the Marconi system. This communication has always existed. The Bible is full of instances and illustrations. In modern times the authentic experiences of Swedenborg alone would tend to convince the reader. And the vast accumulation of evidence is so great that no argument from details need be entered upon here. Any reader who is not convinced of this has only to make his own researches and to form his own convictions. The aim in these pages is, while assuming the truth of communication, to endeavor to trace out the conditions that render it possible and that also establish its probability, even its certainty. These conditions are two-

fold, — those of the very nature of man himself
and of the interpenetration of the two successive
environments, the physical and the ethereal.
We hold perfectly clear and definite relations
with our friends in the unseen, just as we do with
those in the visible world. The only difference
is that the relations with the unseen are more
intimate, more unfailing, more truly a companionship of spirit. The physical body that died was
a mechanism that transmitted this companionship of spirit but transmitted it imperfectly.
The friend who is in the ethereal, with that more
abounding life, is in a more direct relation to
us here than are our fellow beings on earth.

A vast body of communications, ranging practically over all time, have affirmed the existence of
a realm not unlike our own; of continents, seas,
mountains, lakes, forests, rivers; of cities and
of country; of churches, temples, schools; of
music, of lectures, of art, of the worship of God.
But how, we have questioned, can this be?
Now, if the ether of space has the solidity and
the reality that has been scientifically demonstrated by Sir Oliver Lodge, we understand how

it can be. And if the ethereal world is thus inter-penetrated with our physical world (as vibrations prove), we realize how this world is with us in our very midst. Further, and this, too, is a scientific fact, the ether is so elastic that it transmits the slightest impression made upon it, and thus thought, which is the most potent force in the universe, is instantly transmitted from spirit to spirit; from one who is still physically embodied to one in the ethereal embodiment. Thought is a power of such invincible potency that the kingdoms of the earth are helpless before it. Love is a force of such divine potency that it takes the wings of the morning and darts, straight as a beam of light, to him to whom it is sent. Thought and love, they are the irresistible powers of life.

The rationale of the change we call death reveals it as no evil, no calamity, but a step onward in our great evolutionary progress. In our more spiritualized religious faith we shall come to recognize death as a sacred festival rather than as an occasion for gloom or sadness. Jesus came to bring life and immortality to light; to demonstrate to us that spiritual life is eternal in its nature. We simply discard successive environments as we go on from glory to glory. Now and here, man, as a spiritual being, has the spiritual organs of sight, hearing, and, indeed, entire perception of presences that his physical eye cannot see. And why? It is very simple. It is a mere technical matter.

In the infinite octaves of vibration, the physical organs of the eye and ear only register a small proportion. Ultra-violet light, for instance (which, in technical language, only begins with the fifty-first octave, and which is demonstrated in the laboratory), is in a vibration beyond that which the eye can register. We recognize here but the smallest proportion of the etheric vibrations. Now the ethereal body is in this state of high vibration and is thus beyond the point which the eye registers. The friend in the unseen stands by our side and we do not see him. In the law of vibration lies the scientific explanation, — an explanation likewise applicable as to why we do not hear his voice when he speaks to us. But there are other ways of hearing than by the

ear. Telepathy is the language of the spirit.
Thought to thought responds unerringly. And,
as is well known, there are the phenomena of
clairvoyance and clairaudience. When man
more fully develops the organs of his spiritual
body, these will cease to be phenomena. They
will be the natural faculties of his daily experience.
"Within, beyond, the world of ether," said
Frederic W. H. Myers, "must lie the world of
spiritual life. That the world of spiritual life
does not depend upon the existence of the material
world I hold as now proved by actual evidence.
That it is in some way continuous with the world
of ether I can well suppose."

This is to say that Mr. Myers, in contemplating
the cosmos, recognizes as its first three states the
physical, the ethereal, and the spiritual. Each condition is natural. There are no startling and revolutionary changes. There is no lapse of consciousness. The absolute continuity of consciousness is
the truth at the very foundation of our spiritual life.

We need to disassociate the idea of our *life*
from that of the duration of our *physical* life.
Whether in the physical body and environment, or

in the ethereal body and environment is immaterial, just as one's changes of costume are immaterial to his essential life and pursuits.

The continuity of consciousness is as unbroken and as uninterrupted by the withdrawal from the physical mechanism as is the consciousness and the power of the musician by the loss of his piano or violin.

The Gates of New Life are thrown open to the man who has passed from the physical to the ethereal worlds. It is all so natural to him that many persons, indeed, have to be convinced, that they have made the Adventure Beautiful. Doctor William James is one who has said that he had to be led to look upon his physical body, as it lay on the bed, before he could believe that he had passed on. In a communication received from William T. Stead (three days after the *Titanic* had gone down and two days before the arrival of the *Carpathia* in New York had brought tidings of certainty to any one), Mr. Stead, as recorded in another book of mine,[1]

[1] "The Adventure Beautiful." Boston. Little, Brown, and Company, 1917.

stated through the hand of a friend (who was not a professional psychic) that his dead son met him and assured him that he had passed into the next phase of life; that he too was what we have called "dead." Continuing his automatic writing Mr. Stead added: "I looked down at myself; I looked as I always had; and I said, 'Oh, no, this cannot be true.'" The remainder of the story, which I will not entirely reproduce here, was not only intensely interesting, but a narration to throw much light on the conditions beyond.

Still more convincing is the instance, recorded in the same book, of the transition, and subsequent message regarding it, of Mrs. Sylvester Baxter (Lucia Millet, a sister of the well-known artist, Frank D. Millet), because the message from Mrs. Baxter included such verifiable matters as to be unmistakably evidential, even to the most sceptically searching inquiry. An early experience of my own, occurring at sea, on the night of May 19, 1896, has always persisted in vivid memory. It was this:

Wakened in the night by what seemed a current of electrical shock, I seemed to know (rather than see) that three figures stood near with an indescribable sense of joy and surprise; and the words, "Is *this* all? It is all over!" that (by some inner perception) I also seemed to know rather, even, than audibly to hear, were spoken by one who had just passed into the ethereal. Afterward I learned that this was the date coincident with the death of Kate Field. Some months later when, by the arrangement of Doctor Richard Hodgson, I had a series of séances with Mrs. Piper, Miss Field being the chief communicator, I asked her, at one time, to describe to me just what happened on her first consciousness of having withdrawn from the physical world. "I found myself standing on the floor," she said, "in the room in which they had laid my body on a long table. My mother stood by me, and said : 'Kate, my child, have no fear; come with me.' And she took me to the house where were my father and my brother." In this connection Miss Field also said that in these first moments she thought of me, and that her mother told her she would show her the way to find me. My experience that night on ship-board was described through the automatic writing by Mrs. Piper's hand; although at that time it had never been made known.

The general consensus of testimony is as to the absolute naturalness of the experiences on entering the Gates of New Life. The friends who have been known and loved on earth, and who have already passed on, meet the one newly arrived and explain and assist in the adjustment of the new conditions. To a preponderating degree the testimony is that almost the first thought and desire is to be able to make some sign or token to those left desolate on earth: to assure them of the perfect continuation of life and love. The success in conveying this assurance rests with us as much as with them. If we are unable to respond to these higher vibrations of touch or tone or thought, they have no miraculous power to impress us with these manifestations. It must always be, for the most part, a spiritual recognition, and not any expectancy of physical phenomena. The highest order of communion between two is when both meet in aspiration and love and the nobler activities. There is no union of spirit comparable to the uniting for a noble purpose. Instead of that grief which saddens and pains those so infinitely dear, let the one left on earth enter on some special line of sympathetic and helpful work and call on the friend in the unseen to lend a hand. It will be amazing to see how difficulties are smoothed away; how circumstances will be adjusted; how one will be prompted to take the right path, to meet the right person, to find the right book, — to be led through experiences which, while all natural, yet still combine to form a mosaic of complete preparation, or which further the achievement of the purpose in hand. The spiritual world is an inclusive phrase; it includes the present, in a discrete degree, as surely as the period beyond. To live the life of the spirit is to live in the spiritual world, whether here or hereafter.

The interlude of friendship and companionship that exists during the period when one of the two who made up life for each other is in the ethereal and the other here may be made one of ineffable blessedness. It rests with ourselves to make it so. In the almost universal bereavements in this War a great opportunity is offered for entering into a higher spiritual consciousness. We best learn the divineness of life by entering into the divine realm. And this realm is open to each and all of us, at any moment. It is the realm of high and beautiful thought.

"Blessed are the songful of soul;
They carry light and joy to shadowed lives."

To enter into the region of beautiful thought is to enter into the heavenly life. We build our own spiritual life, day by day; and thought is the material of which it is wrought. By dwelling on that which is irritating, annoying, sad, or depressing, we deplete our forces. We also create around us an atmosphere impenetrable to the more lofty and beautiful spiritual influences. And more, we injure those we love who are in this realm of thought and beauty. The Gates of New Life are open to all who lift life to the level of unbroken communion with the mystic, in-dwelling Christ. Nor is this mere phrasing.

It is a work; it is a life work. Because the
ordinary life in the physical world is inevitably
full of all possibilities of discord. One does not
need to offer any catalogue of the things just,
or unjust, as may be, that are difficult, depressing,
irritating. No one is free from these. But the
effect they have upon our lives and conduct is
within our own control. A man has been
wronged, misrepresented, defrauded. He may
be absolutely blameless. But the sooner and the
more entirely he can banish it from his memory,
the sooner he can forgive as well as forget, and
the better for his spiritual progress. Sooner or
later he must forgive, for that is the law. Is it
not better to rise to this at once and thus enter
on peace of mind again?

The region entered by the Gates of New Life
is a spiritual region. They who understand and
thus keep to a high order of thought are spiritually
companioned by their beloved who, being free
from the physical discords, are dwelling therein.
Nothing can separate those who inhabit the same
atmosphere of thought.

It is in this natural companionship of spirit

that the most satisfactory communion is found. Meeting Edward Everett Hale soon after the death of his youngest son, Robert Beverly, who had been his most intimate and inseparable companion, Doctor Hale said, reaching out his hand with its warm and generous clasp, "You don't know how well I bear it; Robbie is with me all the time. He walks the streets with me; he sits beside me in my study." By this, Doctor Hale meant the companionship of spiritual perception alone. He was not designating any phenomenal experience. His son was not visible to his physical sight, nor tangible to the touch of hand. But the spirit-to-spirit recognition was unerring. How could it be otherwise when the two were so closely conjoined by love and by temperamental affiliations? The spiritual self, with its increasing development of spiritual faculties, transcends the barrier of the physical encasement. It is the same order of direct communication that might be if two persons, muffled and enveloped in clothing and in masks, who could not see each other because of the covering, were yet side by side and could converse

together, directly, with no difficulty. The spirit
language is evidently not words, but thought,
although this thought is instantly and uncon-
sciously translated into words. The impressions
conveyed are beyond language; yet they are ˏ
translatable into language.

One finds much trace of this order of communion with the invisible world among the Greeks.
Plotinus, whose life on earth fell between 204
and 269, A.D., thus relates an experience:

"Often when I come to myself on awakening
from bodily sleep, and, turning from the outer
world, enter into myself, I behold wondrous
beauty. Then I am sure that I have been conscious of the better part of myself. I live my
true life. I am one with the divine order and
rooted in the divine. I gain the power to transport myself beyond even the super-world. After
thus resting in God, when I descend from spiritual
vision and again form thoughts, I ask myself
how it has happened that I now descend and that
my soul even entered the body at all, since, in
its essence, it has just revealed itself to me?
Man learns about divine things by leading his

soul to know itself as spiritual that it may find its way, as a spirit, into the spiritual world."

Porphyrius, a disciple of Plotinus (born in Syria, 233; died in Rome 304, A.D.), has thus spoken of his inner experiences:

"The soul has the power to extend her activity to any locality she may desire. She is a power which has no limits and each part of her, being independent of special conditions, can be present everywhere, provided she is pure and un-adulterated with matter."

That is to say, the less a man is entangled with materiality, the more clear, direct, and potent are his spiritual power and spiritual perceptions. But let this idea be not misleading. A man is not necessarily entangled with materiality, nor hindered from leading the life of the spirit, because he is dealing with material things. He is in a physical world, and physical matters are his inevitable factors of achievement. The life of the spirit does not mean sinking into vagrancy, idleness, or pauperism. The life of the spirit may be led by the most vigilant laborer; by him who is delving in the mine or laying pavement in the streets; by the man who is controlling vast and intricate industrial interests; who is commanding or serving in armies; who is in office, shop, study, or studio. The life of the spirit does not imply uselessness, but, instead, the highest degree of usefulness and efficiency. For the life of the spirit is in qualities; it is in justice, honesty, consideration, generosity. The man who is at the head of a great railway system, with its vast complication of the human factor and the industrial and commercial responsibilities; the man who is sending ships engaged in international traffic and transit across the ocean; the man who administers the power of carrying on manufactories and industries; as well as the educator, the preacher, the philosopher, has every condition for living the life of the spirit. Let no one imagine that the path to the diviner life and the life of the spirit is in mere inaction; on the contrary, it is the path in which one is charged with the highest energy.

The conception that there is no compatibility between the life dealing with spiritual and that dealing with material things; that the one must be chosen to the exclusion of the other, was the fallacy of medieval times. It was then believed that the life of the spirit was lived by the mendicant; the material life by the producer. It was held that the life of the spirit could only be most truly lived in the seclusion of convent or monastery, while we now realize that the field is the world. Jesus lived no life apart. He went up into the mountains; He sought solitude at times for that unbroken communion of prayer that recharges the spirit with divine magnetism; but he lived his life among men. He shared with them all that they could receive of spiritual riches. Man would not have been placed in a material world if he had not been intended to deal with its conditions. They constitute for him a school of discipline and training. The physical environment is the theater for all possible exercise of spiritual qualities. To become just, truthful, honorable, noble, — under what phase of discipline could man better learn those lessons and develop those powers than just the conditions in which we now find ourselves? But it is our consciousness and our increasing knowledge of the

unseen which conduces to this increasingly higher
life. It is the realization of the unbroken continuity of life that sustains the spirit through dis-
couragements and denials and defeats; that
whispers the truth that these are but temporary;
"just a stuff to try the soul's strength on;" that
defeat and disaster are as valuable in relation to
the wholeness of life as are triumph and prosperity. It is the realization of this unbroken
continuity, the purposes in view not interrupted
by the change of death, that sustains and inspires
human life.

II

THE UNBROKEN CONTINUITY OF EXPERIENCE

"And tears are never for those who die with their
face to the duty done."
—JOHN BOYLE O'REILLY on "Wendell Phillips."

NEVER was there a time when the world
so eagerly questioned about the nature
of the next phase of life as now, when
these untold thousands of our youth have suddenly been passing from the battlefield into the
ethereal realm. The research into the superphysical has become an enormous quest. It is
not irrational to believe that this is one of the
results for which the War was here. For, that the
most appalling conflict in all history came upon
us by chance is not a tenable conclusion. Numerous are the reasons assigned, as formulated
by statesmen and moralists.

One writer, in an able analysis of the political
and economic causes for the most appalling trag-

edy that the world has ever known, sums up all
these reasons in one, — "man's failure to live as
God commands." Nor can this be regarded as a
mere phrase of rhetoric. "God's command"
is a law as inescapable as is the law of gravitation. He who breaks it must suffer the penalty.
We find the writer saying:

". . . I have heard the statement that just
previous to the War civilization was at its highest
stage; mankind had evolved — developed, if
you like — to a point never before attained;
education was more general than had been known;
even the spirit of charity was evident in all lands,
among all races; in fact, the world was going very
well and the dawn of a better day was clearly
visible. Therefore, such a climax of horror and
suffering, such a tempest of the brutal instincts
of primitive man, seems to be a negative answer
to man's well-founded hope of a better and a
brighter day. . . . If a few years ago a prophet
had declared what the world would see during
1914–1919, he would have been judged by the
majority of mankind fit for the asylum."

The special command that man has broken is

cited as the law, "Thou shalt love thy neighbor as thyself." Justice and consideration are enjoined; but selfishness has largely ruled. Now if the teachings of Jesus regarding the conduct of human affairs are of any value they are practicable. If they are not practicable, they are of no value. The counsel to love one's neighbor as one's self is not that of a fanatic. It is the counsel of simple justice. Emerson notes that a time comes in a man's development when he is careful that his neighbor shall not cheat him. At a still higher degree of development he is careful that he shall not cheat his neighbor. The student of Emerson finds that he continually affirms the solidarity of society. "It is as great a loss to me that others should be low as that I should be low," we find him saying, "for I must have society." It is an entirely practicable ideal suggested in the counsel of Phillips Brooks : "Be such a man, live such a life, that if every man lived as you do, this earth would be heaven." All these ideals are intimations of a marvelous reality on whose threshold we stand.

It is nothing less than the threshold of an

entirely new comprehension of the nature, the progress, the destiny of human life.

One signal factor in this new initiation has been the service of Frederic W. H. Myers, whose place in the world of letters as a scholar of the finest classical culture, a critical thinker, and a poet, was so widely recognized as to give due prestige to an incident in his life which has led to far-reaching consequences.

It was on the evening of December 3, 1869, that Mr. Myers and Professor Sidgwick were out together for a starlit walk. Mr. Myers was a young man of twenty-six. Of this walk he afterward said to a friend, "I asked Sidgwick almost with trembling whether he thought that when tradition, intuition, metaphysics had failed to solve the riddle of the universe there was still a chance that from any observable phenomena — ghosts, spirits, whatsoever there might be — some valid knowledge might be drawn as to a world unseen. Already, it seemed, he had thought it possible; . . . and from that night onward I resolved to pursue this quest." Thus was initiated, in that one moment, the signal pur-

pose of his life. Mr. Myers held the conviction that if a spiritual world ever had been
manifested to man it must be manifest in the
present just the same. He more or less clearly
perceived that the entire life, the energy, of every
day depended upon some influence from the unseen. Was there in man "an abiding psychical
entity far more extensive than he knows, — an
individuality which can never express itself completely through any corporeal manifestation"?
Could the spiritual man function separately from
his physical body? Was the real personality
capable of being liberated from its material organism? Was there truth to reward him who
should diligently search in the mysterious realms
of occult phenomena? Was the man, the spiritual man, in reality independent of his physical organism? Nothing less than this was the
sublime quest on which Frederic Myers set out
from that night. When (on January 17, 1901, in
Rome) he passed into the unseen, did he find the
answer to his life's questioning? The little tablet
placed to his memory in the English cemetery
in the Eternal City, forever poetically consecrated

by the ashes of Keats and Shelley, bears this
fitting inscription : "He asked life of Thee, and
Thou gavest him long life ever and forever."

At all events Myers dedicated his life, his
genius, to this inquiry. Flournoy well says of
the spiritistic doctrine of Myers, "If future discoveries confirm his thesis of the intervention of
the discarnate in the web and woof of our mental
and physical worlds, then will his name be in-
scribed in the golden book of the initiated, and,
joined to those of Copernicus and Darwin, he
will complete the triad of geniuses who have the
most profoundly revolutionized scientific thought,
in the order, Cosmological, Biological, Psycho-
logical."

That epoch-making book, "Human Personality", which Mr. Myers left as his imperishable
legacy to mankind, and which was not published
until after its author had passed from the realm
of questioning to the realm of replies, is an en-
cyclopaedia of the most profound and scientific
investigation of phenomena. It is scientific, it
is philosophic, it is religious. Its depth and sin-
cerity of religious tone impart to its scientific

and philosophical scope an irresistible claim to value. The author studies the problem of telepathy as to whether this is the law of the direct intercommunion of the spiritual man; whether it is a supreme truth, reuniting all beings, — those in the physical realm, those who have withdrawn from that realm, — whether it is the law that unites them all "in a splendid universe of moral and spiritual life"? The problem of the subliminal consciousness; the problems of duty, prayer, life eternal, and all their relations to the life that now is, as well as to that which is to come; the mystery of genius; these, and other vital questions are marvelously discussed in these two large volumes of "Human Personality."

Now life may be defined as the adventure of the spirit into temporary conditions which are ever increasing in significance and enlarging in their horizons; or which decrease in significance and power of satisfaction, and whose horizons narrow instead of enlarge, according to the personal power that is brought to bear upon them. This power is increased or decreased in its nature by the degree of the goodness and intelligence, or of

the evil and the ignorance of the man himself.
For all objective conditions are fluctuating and
are relative to the degree of individual control.
There are certain laws of nature which are fixed,
as the law of gravitation, for instance. In relation to these, man must control his own attitude.
He cannot defy the law without suffering the
penalty, but it is in his power to control his own
attitude in relation to the law. The fluctuating conditions of health, or illness; of some
reasonable degree of success and prosperity, or
failure and privation; the achievement of increasing stores of knowledge, or the remaining
in ignorance, — all these and others which need
not be cited are a part of "the flowing conditions
of life" over which the individual may also exercise an increasing control. Even the momentous
question of immortality (in its differentiation
from merely continued existence) is subject to
the power of the individual. For immortality is
not merely being alive after the change of death;
it is the condition of being alive now! It is a
matter of spiritual vitality. To be just, consider-
ate, sympathetic; to hold service as one of the

priceless privileges; to be generous rather than selfish; responsive rather than indifferent; truthful and noble in every respect, to be active in all that makes for the usefulness and happiness of the largest possible number, to keep one's spirit in sensitive response to the guidance of the Divine Spirit—this is to be immortal in the present. Immortality is not a condition, not a locality. The question is not so much, Shall we be immortal? as it is, Are we immortal at this moment? Immortality is something to be achieved and increased by living in the sympathies and the activities that create immortality. In so much greater measure, then, as one has developed these qualities of the spirit before death, is he the more fitted to enter on this next higher plane of life. "Let this mind be in you that was in Christ Jesus"— that mind that is love, joy, peace, righteousness. To "bear much fruit" in that the Father may be glorified is to live in the widest relations with one's fellow beings; to render the service needed at the moment, not counting the cost; to give the gift that is helpful, though it leave one's own hands empty. For spiritual treasure is infinite,

and to him who lives in the spirit the supply is
sure. Human life is potentially divine life. Religion, in its highest possibilities, is a life and not
a litany, although the litany gives its strength
and support and direction to life.

It could not be assumed that the founding of the
Society for Psychical Research in 1882, some
years after the resolution of Frederic Myers to
devote his life to the quest outlined above, was
in itself the initiation of a new and higher
spirituality of life; but that it has been a contributing cause no one can deny. The last quarter
of the nineteenth century revealed many phases
of new ethical movements. The reconcilement
of science and religion began; they were seen to be
not mutually antagonistic, but complementary
and mutually supporting. Theosophy arose,
offering a great explanation of the phenomena
of the universe; of the problem of the origin,
progress, and destiny of the soul. Spiritualistic
phenomena had opened the way for more from
the mid-century years. Accepted, or denied, it
challenged attention. It became a factor in religious life. All these movements, and the increas-

ing enlightenment of humanity, created a moral preparation for a more highly developed order of human life.

Now here we see the contrast of two great opposing forces advancing towards the future: Germany, with her imperialistic and military ideals teaching the doctrine that Might, not Right, is the arbiter of national destinies; England, France, America, Italy, and other nations imbued with a purer ethical purpose. How could the advance of two such utterly opposite movements, — the one for physical domination, the other for moral and spiritual domination, — result in anything else than a terrible conflict?

For what was this War? Had it not aspects unknown to the historic past, and that brand it as a new order of human tragedy? "For we wrestle not against flesh and blood, but against principalities, against powers, against the rulers of the darkness of this world, against spiritual wickedness in high places." Then what remains? What can we do?

In this War we encountered not men; we encountered fiends from Hades. The editor of a leading American journal thus characterizes the Prussian policy:

" 'The enemy must be thoroughly engaged at once.' Nothing could better illustrate the nauseating hypocrisy, the bloodless formalism unconvincingly covering a bloodthirsty savagery which so constantly characterizes the Prussian beast. Who are 'the enemy' that are to be 'thoroughly engaged'? Are they fighting men who can fight back? Not a bit of it. They are unarmed, non-combatant messengers of mercy — ambulance men risking their lives in the always perilous No Man's Land that they may perhaps ease the pain or save the life of some tortured and helpless human being ripped open by shrapnel or left with a bullet-shattered limb, suffering through terrible hours the torments of the damned! These heroes of pity, standing right up in the daylight, human targets that cannot be missed, men who have not fired and will not fire a shot in this war, are to be mercilessly mowed down by machine guns. . . .

"So this official order to leave the dead uncoffined and the wounded uncared for comes as no surprise. It is the proper fruit of the upas tree. It is akin to the deliberate and officially ordered bombing of hospitals. It is typical of Prussian militarism. It is precisely the sort of thing that our young men have sailed away across the Atlantic to uproot and finally destroy.

"The German army! What is it in reality? A collection of cowards who shoot down Red Cross men, ruffians who rob and 'beat up' helpless civilians, beasts who mutilate children, criminals who poison wells and even give deadly sweets to babies, torturers who crucify prisoners and abuse wounded enemies.

"Leave the dead unburied! Abandon the wounded to writhe in agony under the burning midsummer sun, without water, without succor, without pity! Shoot down the Red Cross stretcher-parties! These are official German orders. This is the sort of enemy our boys fought in France."

In this startling presentation of the powers of darkness which our young men nobly sprang to overcome is revealed the conditions they met. Then what follows?

"Wherefore take unto you the whole armor of God, that ye may be able to withstand in the evil day and having done all, to stand." For this world is being prepared for the diviner life to come in. The ethical forces had long been gathering new strength and manifesting themselves in new forms of activity; the materialistic and inhuman forces of Prussian militarism had also long been gathering new strength and manifesting themselves in increasing activities. The conflict was inevitable. The Powers of Evil closed in a deadly grapple with the Powers of Good. The Powers of Darkness and the Powers of Light were in their conflict.

It was to this awful combat that the Flower of American youth went forth. The hour is consecrated with their holy knighthood.

"And tears are never for those who die with their face to the duty done!"

The material, the spiritual, were arrayed against each other. It was such a conflict as no age of the world ever witnessed before. For evil forces and righteous forces cannot dwell together. And the reason they cannot longer dwell together in any semblance of peace lies deeper still. It is that humanity itself has now advanced to that degree of spiritual development that requires for its existence and nurture a purer environment. No nation is wholly righteous, or without grave sins against the ideal state. Humanity has developed to that higher degree when it can no longer condone its own sins, whatever they may be. Temperance, economic and social justice, must now come. It is the law and the prophets. History reveals that at intervals of about two thousand years there appears some order of a restatement of spiritual truth; a new manifestation; a new call to "Turn to the Lord and live." For in God alone is life.

> "For Evil, in its nature, is decay,
>   And any hour may blot it all away."

May it not be true that now, at the approach of two thousand years from the appearance of Jesus, the Christ, a new wave of spirituality sweeps over the land? But does so divine a thing as spirituality of life manifest itself in aspects too appalling for reference? The tragedy of Belgium;

of the *Lusitania*, of countless atrocities, are these
the pledge and prophecy of a new wave of spirituality? The association of the two is unthinkable
and incredible. So we might rationally say.
There is a mystery still deeper than this. May we
try to penetrate it, in however feeble a measure?

It is an established truth that God works
through orderly laws. Evolution, not revolu-
tion, rules the kingdom of nature. If we sow
wheat we do not reap a harvest of tares. Cause
and effect go hand in hand in orderly sequence.
But the very advent of a higher wave of spirituality forces a deadly conflict with the evil that is
in the world, both individually and nationally.
If a man to-day rise to a new height of spiritual
power, what is the first effect? It is to extermi-
nate the sin that he had yesterday. If he were
unjust yesterday he must free himself from injustice to-day. Now the very degree of moral
development that humanity has achieved will no
longer tolerate the sins that civilization, up to this
time, *has* tolerated. The very good focuses the
evil. The conflict was inevitable. The causes
had existed in the immaterial world. They were

recognized by the spiritual consciousness of mankind. They encountered the invisible challenge of this higher moral consciousness, and they crystallized and formulated themselves for the awful conflict.

On the higher plane the War was a spiritual drama. We have talked of Armageddon; we saw it before us. The Forces of Good, the Forces of Evil, met in their grapple. Now, in relation to the youth who have leaped forward into this conflict; whose noble purpose, whose high enthusiasm, whose devotion to lofty ideals have led them on, — what is revealed to us when they sacrifice their physical life in this tragic struggle?

This is revealed : that these gallant young spirits have forever allied themselves with all that makes for righteousness; that their devotion to true ideals has consecrated itself by seal and sign eternal!

They have died that the noblest ideals of humanity shall live! What ineffable blessedness is theirs! What ineffable blessedness is ours by all that sharing of their nobleness through undying love!

Humanity has now achieved that degree of

spiritual development which requires a finer and purer environment. That is what this War, effacing and exterminating old conditions and creating new ones, is to give us. A world remade beckons us on in a not remote future. It will not be a sudden transformation. We shall not close our eyes in sleep on the world as it is and awaken in the morning to find it transformed to paradise. But that we are at that standpoint, even now, when all conditions for life are contemplated from a loftier range of vision and estimated by purer ideals, could hardly be denied. The larger recognition of the spiritual forces of life in the scale of the practicable and the applicable is, in itself, a signal advance of the race. It is not the lack of sound judgment, but the test and the sign of the sound and wise judgment to recognize unseen forces as those whose influences are the determining and the permanent. The hardships of the physical life increase; physical resources constantly become more difficult to compass. What then? Are we to learn that beyond the physical, — in the superphysical realm, — exists an infinite supply on which, hitherto, man has drawn to only a very slight extent? Are we to recognize that when Emerson said, in reply to an assertion that the world was coming to an end, that he "could get along without it", the remark is not mere wit and persiflage, but states an wholly practicable truth? We relinquish the physical resources of life to an increasing extent. They grow more difficult, more impossible for us to compass. The high and ever higher prices of food, clothing, shelter, — the three primary necessities of life, — suggest to one the wonder as to how he is to continue on this planet at all! Travel becomes so expensive that he vaguely contemplates his restriction to such portion of the earth's surface as he may be able to traverse on his two feet. What is to be the end? Are we to be crowded off the earth altogether?

This brings us to the verge of the recognition of the true nature of our life.

Man is a spiritual being and an inhabitant of the spiritual universe. It is only in the most temporary and fragmentary sense that he is a physical being and an inhabitant of the physical universe. His nature is so largely adjusted to respond to higher realms that the fact of being, as it were, compelled to transfer much of his life, here and now, to those higher realms, cannot be a misfortune. It is as if he were inhabiting only the lower floor of his dwelling, while above were successive floors far more delightful. But he remains on the accustomed level and will not be persuaded to mount higher. Suddenly floods come; or fire invades his familiar interior, and to escape destruction he must ascend to the next story of his house. Once bestowing himself there he finds it far more desirable; but he would never have made this change had he not been forced into it. Is it not possible that this analogy explains the present condition of humanity? Are we not being forced to a higher level of life? Our real world is that among the unseen potencies and under superphysical conditions.

In the middle of the nineteenth century people were crossing the continent to the Pacific coast in conveyances drawn by horses. A quarter of a century later they were crossing it in railway trains. The steam engine had taken its own place. Morse invented the telegraph which carried messages with a rapidity undreamed of before. Marconi perfected the system of sending messages through the ether. We have learned to navigate the air and to sail under the surface of the water. The horse is superseded by the motor car. Entering into the use of the more subtle mechanical forces, man will also develop and use the more spiritual forces in application to his personal life. Immortality is more in increasing degrees of consciousness than it is the question of duration. He who lives in a more abounding spiritual consciousness, now and here, is thereby more immortal. For in consciousness is the true life.

And then? Then it is for us, for those in the seen and in the unseen, to unite in building a new world. If the War leaves us no better than it found us, all its appalling tragedy and suffering and its incalculable loss will have been in vain. Are we to take up life again on no higher round? Not so. The evolution of a nobler civilization is working itself out on lines of harmony with the eternal purpose. All the ease and pleasure and joyfulness of life that seemed so innocent and so full of enjoyment was yet deteriorating if it tended

to retard this nobler progress into the new civilization; if we rested content in it, knowing how
imperfect was its structure; knowing that it har-
bored economic injustice, selfishness, self-indulgence; that it tolerated sins of omission and
commission. Yet it was a pleasant, easy-going
life, with an abundance of charity, even if not
over-abundant in justice; not without its nobler
aims, even with rather prevailing ideals of having
a good time. For the most part all fairly well-to-
do people had a very good time, indeed. In the
old, easy-going sense of those days, no one has a
good time now. Those good times were not, in
themselves, evil, but *if* they were retarding the
more noble organization of society, then they
should give way to these more difficult conditions
which are yet doing the nobler work in forcing a
more just and a finer adjustment of national life.
Not unfrequently is destruction the initial step
toward regeneration.

Two forces are now in mortal combat; one is
evolving the divine harmony; one is opposing and
retarding that evolution. What service is being
rendered by this retarding agency? It is within

the personal choice of each man to identify himself with that which is advancing all that is noblest
in life, or with that force which is opposing it.

"See, I have set before thee this day life and
good and death and evil; . . . I call heaven and
earth to record this day against you, that I have
set before you life and death, blessing and cursing;
therefore choose life. . . . That thou mayest
love the Lord thy God, and that thou mayest
obey his voice, and that thou mayest cleave unto
him; for he is thy life, and the length of thy
days; . . ."

And again:

"Be strong and of a good courage, fear not, nor
be afraid of them: for the Lord thy God, he it is
that doth go with thee; he will not fail thee, nor
forsake thee."

To identify one's self with the forces of higher
progress which are those of the life and good, as
against those which are for death and evil is to
go on in an unbroken continuity of experience,
whether in the body or out of the body. The
spiritual man has thus identified himself with that
which is permanent and immortal.

"... What is excellent,
  As God lives, is permanent."

For it is that which is —

"Built of tears and sacred flames,
  And virtue reaching to its aims;
  Built of furtherance and pursuing;
  Not of spent deeds, but of doing."

To the soul that has chosen life and good, that has identified itself with the highest, the chrism of the divinest joy is given. It has been finely said of our soldiers that they died that the nation might live. But beyond this is an even greater truth, — they died that they themselves might live! That they thus attained to a life so far more abundant than that which they have laid down that their joy is full.

"Never were there so many knights, or so noble," we find Doctor Stires again saying; "but all grateful for the honor of serving, and all ready to conquer death with a shout or a smile, and gladly to cross the frontier for the higher service. It is light, light, everywhere light, and no darkness at all."

These who make the Adventure Beautiful have thereby so made themselves a component part of the nobler order of life that in this brief time they have thus compassed the spiritual development ordinarily only achieved through a long period of discipline. We can only hold fast to our invincible faith in God. The "dreams of the joy of dear ones in the Life eternal", as so tenderly phrased in some preceding citation from Doctor Stires, are, in reality, spiritual insights and spiritual visions. They are glimpses into the divine realities which God permits us to enjoy for our sustaining and our courage to still press on. Nor are these visions in a merely symbolic sense. Actual knowledge of those in the unseen is wholly possible. Actual communion with them, spirit to spirit, may be enjoyed. Love is the supreme and irresistible potency, and where love unites, all the powers of earth and air are powerless to divide those who are thus united.

The release of the spiritual man from his physical body is not to uncomprehended conditions. Science gives us definite knowledge of the ethereal environment. Consciousness is not a function of the physical brain, but a function that manifests itself *by means* of the physical brain, although it is as independent of this instrument and as much greater than can be thus manifested, as the musician is independent of his piano or violin; or as his resources of music to manifest are as far greater than any instrument can afford him adequate scope for producing. The question of the order of life immediately succeeding the life on earth is a much larger one than that involving the fact of communication alone. It demands a more adequate comprehension of the very nature of life itself. Sir Oliver Lodge says of death that it is "an important and momentous event, truly, even as birth is; a waking up to new conditions, like a more thorough emigration than can be taken on a planet; but no destruction, no lessening of power. Rather an enhancement of existence, an awakening from this earthly dream, a casting off of the trammels of the flesh, the realization of a body more adapted to the needs of an emancipated spirit, the entering on a wider field of service, the uniting with the many who have gone on before."

Communication between the two states is no longer to be regarded as either apart from the religions life, or as chiefly identified with scientific investigation; but as a natural aspect of the inter-relations. For the joys of companionship are not ended with the passing of one into the life beyond; a new order of companionship may be established, with its ineffable sweetness and satisfaction and inspiring joy.-

III

EVIDENTIAL COMMUNICATION AND PROOF

"I transport myself to your side and say, speaking just as you would to any friend, 'Come, I have something to say to you.' I insist until you fairly hear my voice. The flesh is stubborn, and it is often almost impossible to make myself heard. . . . All space is peopled with spiritual beings. When you leave the body you enter this space (as you call it) but which is more solid than a million earths, and all the planets of the universe are but as a pebble in comparison. Death has a great work to perform. Every plan, every movement, is directed from this side. All the discoveries, all the new inventions, are projected from here. Our surroundings are adapted to our uses. We have homes and houses and gardens and streets; but there are mysteries here beyond your power to comprehend. As one rises from realm to realm all things become grander and more beautiful."

COMMUNICATION between those in the unseen and in the seen is so abundantly proven that from this time on, in all discussion of the matter in these pages, it will be taken for granted. If the modern evidence that has accumulated in such vast volume within

the past sixty years, to say nothing of the records
of the Bible and of the entire world, indeed,
from all earliest time, — if all this evidence has
not established its existence, the offering of any
additional matter would be useless. Communication is as well attested as is the working of
the telegraph. Its experience in some form is
an almost universal one. These experiences
occur to those who believe and to those who do
not believe.

The invisible world penetrates the visible,
and throngs of beings we do not see surround us
constantly. The reason we do not see them is
because the etheric body is in a state of too high
vibration to be registered by the physical eye.
In another book[1] I have endeavored to present
the scientific explanation of this in full detail.
In the two chapters in that book, "The Powers
of the Ethereal Body", and "The Nature of the
Ethereal World", it was the aim to make this
clearly comprehensible from the basis of actual
laboratory experiments and from the latest scien-

[1] "The Adventure Beautiful." Boston. Little,
Brown, and Company, 1917.

tific data as evolved by psychic research. In a word, as has already been said, the physical eye and the physical ear respond to only a limited range of vibration; and all that is above or below that range cannot therefore be either seen or heard. The extension of sight by means of the telescope above the range of the eye, or by the microscope below its range, will readily occur to all. Thus those who have withdrawn from the physical plane may be about us, although their presence is not reported by the senses. Clairvoyants claim that they encounter in the public streets as many inhabitants of the ethereal world as of this. The psychology of the future must take cognizance of the development of spiritual perceptions as they become factors in all present experience. The organic spiritual body that pervades the physical body, that has corresponding organs and powers, must be reckoned with. Death is merely the process of separation between these two bodies. The testimony of the senses in regard to a vast range of life is so restricted and limited as to be worthless. Who has ever seen or touched electricity?

What could be the testimony of the eye, unaided by the telescope and the spectroscope, regarding the sidereal system? Epes Sargent, one of the most notable thinkers of the nineteenth century, after presenting a long and convincing array of evidence for the existence and recognition of the ethereal (or spirit) body, says:

"From the facts here brought together, it may be inferred that the spirit body is not a mere hypothesis; it is proved by the phenomena and the inductions of evidence; by the objective appearance of spiritual beings; by the testimony of clairvoyants who can see them, and by the testimony of spiritual beings themselves, who claim not only a super-ethereal organism, human in its form, but the power of assuming visible bodies like those which at different stages of the earth life they had while here; by the phenomena of somnambulism and clairvoyance giving evidence of spiritual senses, for as the bodily senses imply their object, so do the spiritual senses imply *theirs*, and are prophecies of an endless life; by all the analogies that reason and experience supply; and by the belief of men in all ages and climes, — a belief founded on the actual reappearance of those who have died.

"Add to these considerations the facts of a manifold consciousness pointing to a complex but unique organism; also the marvels of memory, in which faulty impressions inhere and persist which are inexplicable under the theory of materialism, involving as it does a constant flux and removal of the molecules of the organs of thought. Only the existence of a spiritual body can account for these things."

As a matter of fact the War, thus precipitating such an enormous number into the next phase of life, compels consideration of their immediate conditions and of their relations to the visible world. The psychical experiences connected with the War are already numerous.

Recently Mrs. D. Parker, of Herts, England, was engaged in some household duty when suddenly she heard her son's voice calling "Mother", as if in great pain. The son was a private in a Middlesex regiment. So real was the voice that she dropped her work and hastened down-stairs, feeling that he must have arrived. The call was repeated, but she found no one. His letters ceased, and she felt as sure that he had passed from this life as she did after receiving, some days later, a notification from the War Office that he had been missing since April 24th, the date on which she had heard the voice. For a time no. word reached her; then a neighbor received a letter from another soldier saying that an Australian battalion had found the dead body of young Parker and had given it a military funeral and burial.

A young American lady, Miss Annie Halderman of New York City, was in London in the winter of 1915–1916, and was one of many of the noble women who "adopted" a soldier in the ranks for whom to personally care. Miss Halderman's charge was a Belgian, and later he was killed on the field. After her return to New York Miss Halderman (whose own beauty of life is an ideal of womanhood) still kept in communication with his wife, who had been left with young children and to whom the sympathy and care given to the dead father was continued. By associating

herself with one or two other friends Miss Halderman was enabled to assure the widow continued
aid that the children might be educated and
cared for.   One night she was awakened by the
feeling of a presence, and in the darkness there
came before her distinctly the face of a man which
remained visible long enough for her to perfectly
see and remember the countenance.   A little
while after, the widow, in a letter of gratitude,
inclosed a photograph of her dead husband,
saying she felt that he would be glad that Miss
Halderman should have it.   It was the face
that had appeared to her!

This occurrence seems to indicate that he
fully understood the aid that was being extended
to his wife and children; that he wished the kind
and generous friend to know that he was aware
of it; that he was in some way enabled to make
his face visible to her, and that he influenced
his wife to send the photograph that she might
identify the face she had seen.

As a matter of fact, the relations between the
inhabitants of the two realms are far more simple
and natural than has been fully realized.   There

is no such separation as is often believed. Nor is communication limited to that which is strikingly supernormal. There is, without doubt, a very large body of communication that is seldom recognized as such because it comes in so entirely natural a manner. It comes into one's mind, so to speak, and is either accepted as one's own individual thought, or as coming from some unformulated source. And it is also true that one cannot prove, even to himself, in many of these cases, whether the matter is, or is not, generated by his own mind. But there are also many cases when the thought, the prompting, or the information so links itself with objective things, unknown at the time to the individual, that he can identify the communication as coming from some one in the unseen and often can even identify the source from whence it comes. Such an instance as this is related by Emma Hardinge Britten, of England, whose initial essays in the world of effort were on the musical and dramatic stage, but whose native psychic gift came to so dominate her that she became an eminent medium. Born in affluence and cul-

ture, Emma Hardinge found herself, in early
girlhood, left, at the death of her father, without resources, and she, with her mother, came to
New York. During the voyage they came to
know one of the officers of the ship, who offered,
on his next crossing, to bring to Miss Hardinge a
package that an English friend desired to send.
The time came when the steamer would have
been approximately due, but no alarm was felt
at a little delay, as the sailing was in the winter,
and ships at that time were frequently some days
late if they encountered severe storms. But one
evening she felt the presence of some one unseen
whom she seémed to recognize intuitively as
this young officer; and it came into her mind
that the ship had gone down and that all on
board were lost. There was nothing visible nor
audible; but to the inner sense all this seemed
to be made clear. She even felt a sensation as
of icy water. Yet nothing that could be classed
as phenomena occurred. The information was
not conveyed with the definiteness of the clair-
audient voice, or of automatic writing. But,
as a matter of fact, the ship was never heard from.

There were no "S.O.S." calls possible in those days. That she went down with all on board the unbroken silence alone attested. It does not require a faith that degenerates into credulity to fully accept the apparent happening that the officer came to Miss Hardinge and communicated to her his fate.

A remarkable instance of communication is related by George Thompson, M.P., of London. Mr. Thompson, recognized as an eloquent speaker in Parliament, came to this country as an anti-slavery speaker, in the decade of 1850–1860. At one time he was the guest of Isaac Post, who, with his family, had been much interested in the spiritualistic phenomena produced through the medium of the Fox sisters, and through whose hand was automatically written the book entitled "Light from the Spirit-World." At the invitation of Mr. Post, Mr. Thompson had a séance with the eldest of the sisters. Some years before this Mr. Thompson had been in Hindustan on a government commission and had made some personal friends among the Hindoos, two or three of whom had since passed to the beyond.

It occurred to him that if he could get a message from any one of these it would be a real test. He mentally inquired if any of them were present, and three affirmative raps followed. His request for a message was also answered in the same way, and the signal was given for using the alphabet. This was a tedious process, but one that was much employed in the early days of messages; it consisted of repeating the alphabet until the signal of a rap indicated the right letter, and thus words were spelled. Mr. Thompson began repeating the letters and received the first signal at the letter "d", followed by the letters "w-a-r-k-a-n-t-h-t-a-g-o-r-e-e." Mr. Post remarked that this was a totally meaningless medley, and that there must be some mistake. He advised his friend to try again. Mr. Thompson studied the slip of paper on which he had written down these apparently unconnected letters, and then exclaimed "Dwarkanth Tagoree!" For here was the Hindoo name in full. Mr. Thompson uttered some friendly words of surprise and delight, to which a shower of raps responded. Tagoree had been a friend especially prized; a man of unusual ability and goodness and also a Hindoo of high rank. By means of the tedious, yet reasonably direct process of the alphabet, a conversation of some half hour's duration ensued. Mr. Thompson put some questions to test the alleged identity. One of these was as to a gift sent by the Hindoo friend to Mr. Thompson's wife. The correct answer (a cashmere shawl) was spelled out. The Hindoo had visited London, and Mr. Thompson asked for the place they had last met? The reply named the place correctly (Regent Street), and one or two other test questions met an equally true reply.

The "Undiscovered Country" is no longer undiscovered or unexplored. But its true nature is only recognized through spiritual perceptions and aspirations. An interesting editorial article in the New York *Tribune* for August 4, 1918, conveyed a surprised but yet enforced recognition of the rapidly increasing interest and belief in the realities of communication between the two realms. The writer, however, instanced Eusabia Palladino's phenomena as something

so remote from any spirituality of life, any true
religious feeling, as to discredit the growing
interest. Now, as a matter of fact, nothing is
less connected with the persistence of loves and
friendships and spiritual intercourse between
those who have passed on and those here than the
crude material phenomena of which the Nea-
politan peasant woman was a striking purveyor.
If it had its own interest in suggesting unex-
plained forces of nature or laws not yet grasped,
that alone might give it claim to scientific investigation.

A still more interesting and remarkable phase
of unquestioned physical phenomena is that so
ably studied and described by Doctor . Crawford in the Irish family, where every opportunity
was gladly afforded him to investigate strange
occurrences. For instance, when a large table
was raised in the air by some invisible means,
Doctor Crawford found that if he passed between
the medium (a young girl) and the table when it
was suspended in the air, it immediately fell.
He set himself to work to penetrate the reason
for this. His investigations led him to con-

clude that some power, like that of a rod, projected itself from the body of the medium and
raised the table; and that his passing between
the girl and the table broke this current of power.
Doctor Crawford's study of this case was carried on with scientific appliances, scales, mirrors,
and phonographs to record and establish the
reality of sounds or raps; and he, as a scientific
engineer, brought to his task trained knowledge
in an exceptional manner. Now, however curious are these phenomena, they are no more spirituality, they are no more religious growth and
culture, than are the experiments in a chemical
laboratory. Persons who should mistake these
for religious spiritualism would go very far
astray. With Eusabia Palladino, when the
exhibition of her powers was given in this country, Doctor Hyslop refused to have anything to
do. Not being a physicist, he was not a specialist
in investigating physical phenomena, and even
admitting its genuineness, partially or wholly,
as may be, it had too little significance for him
to command his time or interest. We need to
discriminate between a possible communion of

spirit to spirit, in all the beauty of love, all the
sacredness of religious feeling, all the recognition
of the communion as natural to the continuity
of life and as simply the continuation of that
spiritual intercourse between the seen and the
unseen that pervades all the Scriptures, — we need
to discriminate between this and mere physical
phenomena, however strange that may be as
estimated from known physical laws.

Let one take some such communication, for
instance, as that received through automatic
writing by Mrs. Fanny H. Park, of Liverpool,
who (under her maiden name of "F. Heslop")
has published, in a book entitled "Speaking
Across the Border-Line", many of these beautiful and most interesting messages received from
her husband. A little word about him contrib-
utes to the better understanding of the messages. John Park was a Scotsman, filled with
the love of life, a keen sportsman, a lover of nature who "revelled in the beauty of river and
loch", and whose bias of mind, Mrs. Park tells
us, "was toward the practical rather than the
poetical, while for mysticism and all occult

matters he had no toleration." Mr. Park was a man of strong affections and tenacious friendships; many of his friends said to Mrs. Park after his passing that he "was the most lovable man" they had ever known. His wife says of him that his character "was a combination of strength and tenderness, strong in rectitude and every manly virtue, but tender and understanding toward the weakness of others." Mrs. Park adds:

"We never spoke of his approaching death, and the thought of his return from the spirit world and the possibility of communion with him never entered our minds. To us, death meant separation, and separation meant death. So when he left me, I seemed in my loneliness and desolation to have passed also into the land of shadows."

As a matter of fact, also, Mr. Park had been intolerant of the idea of spirit communion. Neither he nor his wife felt any sympathy with the theory. But after his death, through the hand of another person, these messages to his wife began to be given through the medium of

automatic writing; they established his identity
so unmistakably that she had no choice but to
accept them. This was rather perplexing to
many of their friends; and he, apparently hearing a discussion that took place, thus referred to
it to Mrs. Park:

"Our friend is quite right in thinking that
when on earth I opposed all suggestion of spirit
communion. I thought there was blasphemy
in the very idea. My whole early training had
bent my mind in the wrong direction. Now,
with my fuller vision, and stripped of all the
theological misconceptions of my youth, I see
how utterly wrong I was. And to me, one of
the most wonderful discoveries of this life here
is that it is possible to return to full communication with you, my beloved, and continue in
almost perfect and unbroken joy the union con-
summated twenty years ago."

Later, he began to use his wife's hand for these
communications. Mrs. Park notes that she was
filled with dread lest these were the product of
her own subconscious mind. Perceiving this,
he wrote:

"I see you have been going through a needless distress of mind as to the authorship of these letters. After much reading of modern literature on the subject you have flown to the conclusion that possibly your subconscious mind was impersonating me, and that these letters were not from me at all. My dear, how could you think such a foolish thing? Have I not given you test after test of my identity? Have you not received information beyond your wildest dreams? Surely, you know by this time that it is I who write to you, my love that surrounds you. Never let this doubt stay with you for a moment again. Cast it out of your mind and cling to the definite assurance which I now give you that I am constantly with you, whether you realize it or not, inspiring your mind, smoothing your path, warding off all evil influences, and loving you all the time with a love beyond anything you can dimly imagine."

Mrs. Park had no thought or intention of publishing these messages, feeling they were a sacred part of her private life, but she was constrained to do so for the same reason that Sir Oliver Lodge

felt constrained to give the widest publicity to
the messages received, or which he believed that
he received, from his son, Raymond. In giving
these in full, with a certain admixture that was
sure to be misunderstood by a large number of
readers, Sir Oliver did violence to his own feelings, but he felt he had no right to withhold any
contribution that could throw light on an important subject. Mr. Park, with the wider
vision of the life beyond, urged the publication
of his letters. He saw in them something that
he believed might comfort the sorrowing. When
Mrs. Park decided to do so he wrote:

"Now I am glad to see you are arranging the
letters I have written you from time to time.
They will be especially valuable to the bereaved.
. . . I am glad you are willing to have them
circulated, for it is just what I tell you in these
letters that needs to be known. How love
grows and deepens on this side; how it can be
communicated to those who are in affinity with
one another (when one is still on the earth plane)
and that is the special work of ministering spirits."

In one of the first of these letters Mr. Park

describes his passing to the spirit life. The matter is made so clear and seems to bear such testimony to the naturalness of the transition that, at the risk of unduly quoting, I shall venture to transcribe it.

Mr. Park wrote:

"When I died I simply fell into a state of unconsciousness and was at once taken into my mother's loving care. . . . Gradually the wonders and beauty of this new world unfolded themselves. The loveliness of the trees and flowers, the grandeur of the mountains, the glint of distant lakes seemed familiar, yet all spiritualized. It was some time before I could realize what had happened, and that death had really passed; so I rejoiced, for my suffering on earth had been great. Then spiritual illumination came to me, I developed new powers, and was literally born again. They carried me to my beautiful home, and every flower I loved was there to greet me. Oh, such roses! Would that you could see them too. . . . How can I tell you of this new and beautiful life? . . . I see now that only the germ of truth is taught on earth, overladen with much error. You hardly realize that you have the power to express God in your lives. . . . Remember, you are building your home here all the time you dwell on earth. It is the outer expression of your thought. All spiritual and beautiful thought produces beautiful surroundings. . . . I am busy perfecting our home, but it cannot be completed until you join me. . . . You are never alone . . . but no spirit, however pure and beautiful, must ever come between your soul and God. Because you have given yourself into the divine keeping nothing of any kind can harm you. Banish every vestige of fear from your mind. You are in God's care, and your guides will help to keep evil influences away."

These last lines are especially suggestive, as many persons make an objection to any idea of communication with the unseen, or to the idea of receptivity to influence from those beyond, by saying that they feel all influence should come to us directly from God. In that they are quite right, only is it not always possible, even in this world, to love God more the more we

love our friends, our associates, or the more sympathy and active good will we feel and manifest
to every one?

> "O loved the most, when most I feel
> There is a lower and a higher;"

And again :

> "The love that rose on stronger wings,
> Unpalsied when he met with Death
> Is comrade of the lesser faith
> That sees the course of human things."

That is, the more entirely the soul goes forth
to the divine; the more one "loves God", to
use a common and ever comprehensive expression, the more truly does he love his friends; and
the converse is also true. We do not make the
objection, in this present life, that we cannot,
or should not, love our friends because we love
God. On the contrary, the more deeply any
nature is attuned with the divine, the larger is
the capacity for associations and friendships.
"My friends come to me unsought", said Emerson; "the great God Himself gave them to me."
Why should the love of God and the love of friends

be in any mutual exclusiveness of each other when the friends have passed into the next phase of life? The divine aid is not less if it come through the means of a friend, in the seen or in the unseen.

An instance of communication from the beyond that is one of the most simple and natural as well as impressive, one which has never before been made public, but which I have permission to use here, — is related by Mrs. Bradley, then living in Michigan. The story would lose if its narrative were changed from the simple form in which she herself relates it, and which is thus given in her own words:

"My name is Nellie L. Bradley, and I have lived for twenty-eight years in Muskegon (Michigan), my present home. My husband and I have been devoted lovers for forty-five years, and I am just a cheerful, plain, sunny-tempered woman, never, at any time in my life, a professional medium, or anything of that sort. Nevertheless, I have had some remarkable experiences in that line, one of the strangest of which I will now relate.

"On the first day of the February of 1907 I was sitting by the window sewing, when the voice of my dead sister said: 'Nellie, you must go away, or you will not live many months.' My sister was Mrs. Villa Stowe, who had lived in Grand Rapids. I had always called her 'Darling', for she was my idol, and the bond between us was very close. She had died in the August of 1906. I had been suffering for some time with rheumatism and was perhaps illy able to endure the chill and dampness of the spring. When my husband came in I told him of what my sister had said, and that she had added that the way would be opened. From that moment I began preparing for a journey, although circumstances made it seem extremely difficult, if not impossible for us to leave."

Mrs. Bradley here explained how the undreamed-of arrival of a friend from Duluth combined with other circumstances to enable them to leave at once, and she thus continues:

"Mr. Bradley had bought tickets for Havana, although he did not know why he chose that city, as we had only intended going to Florida. We stopped in Florida, and only then did my husband tell me that he had extended our journey to Cuba. Arriving at Havana we went to the Hotel Tuileries, and a little later we recalled that a young man from our city, Earl Patton of the United States Army, was stationed in that locality, and we went to see him. On returning we found we had taken the wrong car, and looking about to find some one who spoke our language, we noticed a lady in deepest mourning, accompanied by a gentleman, sitting near us. I turned and said to them, smiling, 'Pardon me, but do you speak English?' He replied in the affirmative and added, 'What can I do for you?' We made known our mistake; he directed us aright and expressed the hope that we were pleasantly located, saying that there were delightful rooms in their hotel overlooking the harbor. He wrote the address on a card, and we left the car; but on reaching our hotel we found them waiting to tell us that the rooms of which they had spoken had been taken meantime, but giving us another address equally pleasant, to which we removed that evening.

This casual conversation, with our thanks for their courtesy, was all that passed between us. Nor did we expect to see them again.

"Usually I sleep well; but occasionally there is an exception, and I soon realized that night that, despite fatigue, I should not sleep. A cold wave passed over me and a voice said, 'This is Marie; they called me Sweet Marie from the old song.' I strained my eyes, startled, and although the street light was shining dimly through the shutters, I could see nothing. Nevertheless I felt this sentient presence, and I said: 'I don't know you; what do you want of me?'

"'Oh!' the plaintive young voice answered, 'I want you to take a message to my mother. I have tried, oh so long, and you are the first one I could talk to.' I protested, 'But I don't know your mother,' and she said: 'Oh! yes, you do. Please tell her I cannot be happy while she grieves so deeply; it holds me to the earth.'

"Now this was not a dream. I was never more completely awake and in full consciousness. I asked Marie questions about herself, all of which she answered, telling me that she died four years ago, at the age of twenty-three. Finally I begged her to leave me that I might sleep; and at parting she said: 'My father will take you by the hand and say that you have given him more comfort than any one else.' In the morning I told my husband of the experience, and he remarked that it would be strange if we met these people again and that he should be glad to have an opportunity of asking them if they had such a daughter. But so far as we knew the incident was closed, and we were so engaged with our sightseeing that we almost forgot the matter."

A few nights later, Mrs. Bradley said, her husband proposed that they should go to dine at "Harvey's", and as he spoke a cold wave passed over her. Before she could reply a voice spoke to her inner ear saying, "No, no, please go to the Chinese restaurant; there you will meet my father and mother and dine with them." Mrs. Bradley was so startled she could hardly relate this to her husband; and he at once replied : " Yes, let's see it out ; it would be strange, indeed, if these people were there."

Mrs. Bradley thus resumes the story :

"We started down under the avenue of date palms, in the moonlight, on our way to the Chinese restaurant, and all the way Marie's voice kept sounding beside me. We found it crowded, but seeing two vacant seats at some distance we proceeded toward them, when my husband suddenly grasped my arm and said, in a low tone : 'If there are not those people we met in the car.' A sudden wave of excitement and awe swept over me as the voice of the dead girl again spoke distinctly at my ear, saying, insistently, 'Ask my mother, ask her about Marie.' The lady and gentleman rose at our approach, with a smile of recognition, and begged us to dine with them. In my agitation I at once asked the lady if she knew any one by the name of 'Marie'? She grew deadly pale and dropping her knife, exclaimed, 'Why do you ask? How did you hear that name? Indeed I know ; she was my darling daughter whom we lost four years ago ; we called her Sweet Marie, for the old song.'

My husband then interposed and begged 'we would say no more until after dinner, inviting the gentleman and lady to return with us to our apartment that we might tell them the story." '

The details that Marie had told Mrs. Bradley proved to be correct in every particular, and her parents were deeply affected. On their leave-taking, Mrs. Bradley further states that the gentleman took her hand and repeated exactly the words about the comfort she had given them that Marie had before asserted her father would say.

This little incident illustrates the natural and simple way in which communication from the unseen is interwoven with the ordinary occurrences of daily life. The great error is in regarding communion and companionship between the seen and the unseen as a phenomenal occurrence, rather than as a natural and, to a great extent, a constant experience in daily life. All tendencies to the abnormal are not to be considered as inevitably conjoined with psychical gifts, but rather as due to their abuse, or their absence. The life of the spirit, whether in or withdrawn from the physical body, is a normal life. So far as it varies from the normal, it is simply defective as a spiritual life. The narrations of the mingled life between the inhabitants of the physical and of the ethereal realms persist through all the ages. Boccaccio, in his life of Dante, relates that when the poet died the "Divina Commedia" was found unfinished, and -the manuscript was sent to Can Grande lacking the last thirteen cantos that now appear. The poet's sons, Pietro and Jacobo, were anxiously questioned about the missing cantos, but they knew nothing of them. One night, however, Dante appeared to his son, Jacobo, "his face shining with light, and when the son asked if he were living, replied: 'Yes; but in the true life, not yours.' Then it occurred to Jacobo to ask his father if he had finished his work before he passed to the true life, and if he had, where was the conclusion to be found. To which question came the answer, 'Yes, I completed it'; and then it seemed his father took Jacobo by the hand and led him to the room in which he had lived and, touching a panel in the wall,

said: 'That which you seek is here'; and having said this, he disappeared." And when the
sons looked, the next day, there were the missing cantos. "And in great joy they copied
them," continues Boccaccio, "and sent them to
Messer Cano, and then added them to the imperfect poem; and in this way the work which
had been carried on so many years was finished."

No one can realize the true nature of the present
life until he also realizes the true nature of the
change we call death. Those who pass on are not
asleep. Those who pass on are not removed
into conditions incomprehensible to those here.
They enter, so far as they are fitted, on more
intense activities and a larger range of consciousness, and thus become more alive than is possible
in the limitations of the physical world. The
conviction of immortality and of the eternal
progress of the spirit requires for its completest
atmosphere of growth and its manifestation in
reality the knowledge of the reality of communication between those in the seen and the unseen.
Without this knowledge there may be (and is)
faith in God and faith in immortality as a condi-

tion, vague and ungrasped, but some way, some time, to be recognized as true; but with this knowledge (of the absolute unity of life and the unbroken communication) the faith becomes clear and intelligible, not vague. It becomes an ever-present reality of the immediate hour, sustaining, encouraging, and revealing the practical nature of the divine aid in every hour of life. It assures us we are not left alone. If the religious man, who does not accept the Spiritualists' faith in the communication and the continued companionship between those who have passed on and ourselves — if he asserts his belief and full reliance on the help of God; if he only looks to Jesus for aid — why, that is good; but that faith is not lessened, nor necessarily at all changed, by a little knowledge as to the ways and means by which the Divine Power helps us. "Are they not all ministering spirits?"

Nor do we fully enter into the realities and the nobler possibilities of the present life until we realize that we are, even now and here, inhabitants of both realms. In every achieve-

ment of life we draw upon ethereal forces. The
ethereal realm interwoven with our own is not
a miracle region; it is another phase of nature.
In fact, life here could not exist at all unless it
drew upon the life beyond. There is a perpetual
inflowing of ethereal energy, and if this were
checked, that which we know as the physical
world would cease to exist. The ethereal world
is far more real than is the physical. Stephen
Phillips embodies an absolute fact in the lines:

"I tell you, we are fooled by the eye and the ear;
These organs muffle us from the real world
That lies about us."

The more clearly the vision extends into the
more real world the more power is unlocked to
draw upon for achievements. Then does one
ally himself with the diviner forces. Then does
he learn how to transmute his energy into power.
For energy is not synonymous with power. Energy may be restless and dissipate itself to little
purpose. Power is calm, serene, uninterrupted,
unremitting, and perfects itself in definite achievements. All problems of life are really spiritual

problems. There is no line of demarcation. In the last analysis Love is the only working philosophy of life. Love is light and beauty and power. Love alone, in the larger and higher sense, makes endeavor successful. "Love feels no burden, thinks nothing of trouble, pleads no excuse of impossibility. He that loveth flieth and rejoiceth." He that loveth dwells in that harmonious atmosphere in which there is no waste of energy. The initial condition for any form of worthy achievement is to banish every discordant thought and establish that harmony which rests alone on the basis of universal love and good will. It is when living in this diviner air that communication with those in the unseen becomes easy and a frequent part of the natural experience of every day.

> "Let nothing disturb thee,
>   Nothing affright thee;
>   All things are passing;
>   God never changeth;
>   Patient endurance
>   Attaineth to all things;

Whom God possesseth
In nothing is wanting, —
Alone God sufficeth."

. Nor is the "possession" of God a mere phrase
of abstract and incomprehensible significance.
It is the practical duty of life, and it is the most
practicable of duties. We possess God when
His divine spirit possesses and informs and
dominates our own. Life is a spiritual drama,
and every day's experience may be invested with
a kind of magical enchantment. The enlargement of interests by the extension of thought
and vision into the unseen; by the consciousness of the constant telepathic communion that
may be held with friends there, is the very re-
demption of life from the commonplace and the
trivial to the plane of the significant and the
universal.

IV

THE NATURALNESS OF THE NEXT PHASE OF LIFE

"The soul looketh steadily forward, creating a world before her, leaving a world behind her, and the web of events is the flowing robe in which she is clothed."
— EMERSON.

"This world is not conclusion,
A sequel lies beyond."
—EMILY DICKINSON.

THE absolute naturalness claimed for the next phase of human life is, by a paradox, its most bewildering attribute. The language of the Bible has been taken literally to an overwhelming extent, where it is intended to be only symbolic and figurative. The literal interpretation of this language has been handed down through so many ages, it has been so universally taught, that it is little wonder the world is so generally disposed to accept these ideas. It is not strange that with the symbolic picturing of a state of rest, the suggestion of activities should seem a desecration. Or if the conviction has been inculcated that sleep, poetically invested, is the condition after this phase of life, to endure until some mystical and incomprehensible resurrection takes place; or that a more or less literal acceptance of golden streets, palm branches, and harps possesses the mind, — these ideas, too, being entwined with tender and sacred associations, — it is little wonder that a different philosophy, one involving no break in the continuity of activities, might be regarded as lacking in religious reverence.

Yet a deeper study of the teaching of Jesus will disclose new points of view. Were the more modern conception of spiritual life based on mere phenomena alone, with little heed of the religious feelings, it would naturally repel persons of the higher order. Unless this somewhat different conception of death can be spiritualized and made a vital part of our religious faith, not held as antagonistic to it, the conception will not meet with any universal recognition nor win any universal belief.

But is it not true that religion is, in its very

nature, progressive? Or rather, perhaps, that as man progresses it reveals its truth to him more and more completely? "I have many things to tell you, but you cannot bear them now," we find Jesus saying. In the enlarging conceptions of scientific truth that record themselves, successively, through the ages, generation after generation, we see how the views of nature change; how the attitude and belief of one century, or one generation, is discarded, or greatly changed, by the next. May this not be equally true in regard to the great problem of the origin, the development, the destiny of man? I do not phrase this, "the destiny of the soul", as if the soul were something apart from the man himself. That phrasing is misleading. It belongs to the past, when the conception of man was that of the visible form which possessed, we felt sure, a soul; but of what mysterious nature could not be conjectured. Now we realize the transient aspect of the visible man; we realize that his physical body is no more himself than his clothing is himself; that the real man is simply manifesting himself by means of his physical body as the mechanism, the instrument, of his contact with the physical world.

What is his destiny as an immortal being? We follow him through the physical environment; what next succeeds that? Can we still follow him after he has withdrawn from the physical world? Can we penetrate into the ethereal realm of "the encircling spirit world"? Through all ages this spirit world has been felt; the intimations of immortality are always in the air. Modern spiritualism focused and verified many of these intimations; the purely scientific work of psychical research has contributed valuable aid; but now intuition and increasing spirituality of life are bringing to bear a force of conviction and a larger grasp of knowledge than has before been revealed. Science and spirituality go hand in hand to this end. Science has revealed and formulated the existence of the ethereal world; spirituality recognizes that this ethereal world, in correspondence with the ethereal body in which man is clothed after discarding the physical, is the natural environment for the next phase of this evolutionary prog-

ress we call life. The existence of the etheric
body is now a recognized fact which is no more
denied than is the existence of the physical body.
After this etheric body shall have served its use,
during the sojourn in the ethereal world, it will
be succeeded by a body still finer and more subtle.
But of these future conditions we can now only
speculate; while with the one immediate future
condition we can already formulate much accurate
and positive knowledge.

Here are two realms, the physical and the
ethereal, that interpenetrate each other; the inhabitants of the former withdraw from it and pass
into the latter. The transition effects no immediate change. Nor is the new environment in
any respect so different from the former as to
amaze the newcomer. The greatest surprise,
indeed, is in the realization that the change is so
much less than has been anticipated. There is
a vast amount of evidence already that sub-
stantiates this statement. To the question as to
how one can know that this is reliable evidence
it may be answered that the identity of individuals on the other side has been so unmistakably

established as to give reasonable warrant for its
acceptance as a fact. Now when the identity
is accepted; when the friend making these statements is one on whose truth and judgment reliance could always be placed; and when the
descriptive accounts of the conditions of life in
the ethereal agree with much positive knowledge
gained through actual demonstration in laboratory research, the assertions and statements made
commend themselves to the mind.

Take the case of a communication from Edward
Everett Hale. When Doctor Hale returned to
his Boston home from a visit in Europe would his
friends have doubted any narration of his about
life, or other matters, in London or Paris? Then
why, if his identity as a communicator is established beyond reasonable doubt, should one
doubt any statement of his regarding his present
environment? I may have related in some
previous book the little incident that I beg to
record here, but if so, it is easy for the reader
already familiar with it to turn this page. It
is so typical an illustration of the perfect naturalness of the next environment into which we enter

that I venture the risk of repetition. Soon after
the death of Kate Field, Doctor Hale wrote to me
in Paris, saying, "I did not know Miss Field;
I hope I shall know her." This was in the
summer of 1896; the years went on, and he also
passed into the ethereal. Doctor Hyslop (who
had not known Doctor Hale) was pursuing his
investigations in psychical research through the
remarkable mediumship of Minnie M. Soule, the
famous Boston psychic, and coming to me one day,
some few years after Doctor Hale's death, told
me that Doctor Hale had apparently been at the
séance that morning and had sent a message to
me, although a message that Doctor Hyslop
found quite incomprehensible. It was, "Tell
Lilian Whiting I have met Kate Field, and that
she is the most adventurous spirit I have ever
seen in a feminine body." But link the message
with the letter of years before and how unmistakable is the connection, the message being a
natural sequence to the letter. In the letter he
mentioned that he had not known Miss Field.
When he himself passes on into the same
environment he not unnaturally meets her.

When in this life Miss Field was one of his most appreciative readers and admirers. His convictions on any matter impressed and influenced her. What more natural than their meeting in the new conditions to which both have passed? And Doctor Hale's characterization of her as an "adventurous spirit" is one unusually applicable. The message given somewhere about 1912 is in perfect sequence to the letter in 1896.

Lady Henry Somerset has related that an audible voice out of the unseen spoke to her, directing her to go forward in the temperance movement. At that time she was entirely engaged in the social life that presses upon an English peeress, and while she had felt promptings and drawing toward work of reforms, involving leadership and its sacrifices, these promptings had only dimly stirred in her mind. When the voice spoke her resolution was taken, with the important and beneficent results to the world with which the public is familiar. Nobly did Lady Henry respond to the bidding. She answered the call, and the path on which she then entered has been one of strange contrast to that life of ease and luxury which otherwise would have been her appointed way.

In the very interesting reminiscences [1] of Mrs. Julia Ward Howe given by two of her daughters, there is the account of an experience which she spoke of as "a midnight vision." Mrs. Howe was suddenly awakened by some words falling on her mind, as if from a voice, and in her journal she thus recorded the incident:

. . . "There seemed to be a new, a wondrous, ever-permeating light, the glory of which I cannot attempt to put into human words, — the light of the new-born hope and sympathy — blazing. The source of this light was born of human endeavor. . . . And then I saw the victory. All of evil was gone from the earth. Misery was blotted out. Mankind was emancipated and ready to march forward in a new era of human understanding, of all-encompassing sympathy, and ever-present help, the era of perfect love, of peace passing understanding."

This was in the year 1908; and does it not seem

[1] Julia Ward Howe: 1819–1910. Boston. Houghton Mifflin Company, 1916.

to have been an intimation of the sublime ideal toward which humanity is tending, and of the newness of life for which conditions are being shaped and molded by the recent conflict? Mrs. Howe had never been drawn to any special study of psychical literature or speculative theories. But the eyes that saw the glory of the coming of the Lord were the eyes of vision, and without any especial formulating of specific conviction, her daily life was simply the life of the spirit.

Of Mrs. Browning it was said that she spoke not particularly of religion; her whole life was religion; and similarly it might be said of so exalted a spirit as that of Mrs. Howe, that her entire life, philosophic, poetic, mystic, was the life of perpetual companionship with celestial intelligences.

Mrs. Livermore had given much thought to the writings of Sir Oliver Lodge, Frederic Myers, and others eminent in presenting the philosophy of spirituality, and she had come to the definite conviction of the reality of communication between the two realms. Two letters from her, each narrating a striking psychical experience, have already been published in two previous books of my own ("The Spiritual Significance", 1900; "The Adventure Beautiful", 1917), and in many other of her letters to me allusion and assertion and speculative thought regarding the matter were almost invariably expressed. After her own passing she described, through the hand of a psychic, her joyful entrance to the ethereal, saying in part: "They were all here to meet me; my dear husband, Lucy Stone, Wendell Phillips, and so many of my friends." What more natural? The language used in relating this included many turns of expression characteristic of her, and one or two incidents that corresponded with some objective occurrences, thus establishing a strong presumption of the evidential character of the message.

The etheric double of the individual has its prototype in nature. Every tree, every object manufactured by man, every aspect of nature, has both its material and its ethereal side. Of flowers, we on earth take the material flower; those in the next environment take the ethereal part of the same flower. The material and the ethereal are conjoined like shadow and substance. And, like these, the material corresponds to the shadow; the ethereal to the substance. It is the ethereal which is the positive, the significant, the substantial; it is the material which is the transient and of lesser significance. It is the ethereal body which Saint Paul asserts to be the "substantial" body.

An entire fallacy has been presented and perpetuated under variously erroneous forms. The phase of life succeeding this has been identified with the shadowy, the wraithlike; it has been relegated to a region of phantoms and phantasms; it has been regarded as unknown and, so far as human intelligence could go in the present, as unknowable. Even in the assertion of many of the professional psychical "researchers", the next condition of human life has been presented as something so mysterious that only the scientist should make any attempt to explore it. They would seem to regard it as some abstruse problem in physics or some dangerous experiment in chemistry might be regarded, — as impossible of approach save by the expert. Practically, the attitude of many of them affirms that the general public should provide the funds for carrying on a purely scientific work, whose processes it must not expect to be instructed in, or even hope to understand; and must quietly await results as to whether these experts discover that there is, or is not, personal immortality! As well might the church universal affirm that religion is no affair of the layman; that it consists in mysterious rites known only to the priesthood and exclusively to be directed and carried on by them. The great fallacy has been in relegating the experience entered upon by humanity after the change called death to the region of phenomena. Spiritualism has also largely contributed to this false attitude, although it has contributed so much of truth and illumination that it savors of ingratitude to arraign the movement for its errors. All the same, in the pursuit of truth one knows neither friend nor foe; and there could hardly be found any ethical cult that has not its errors and its abuses. Cults are composed of people, and the human race is not yet infallible; not yet perfect, but simply on its great way toward the goal of ultimate perfection.

The general recognition of the exceptional persons known as psychics, or mediums, has created a widespread (but wholly erroneous) conviction that these persons were the gate keepers, so to speak, and that no communication with those in the unseen was possible save through their agency. Now it is true that there are these exceptional individualities who have the natural gift, in varying degree, of communicating with those who have passed into the ethereal world. Just what qualities or faculties determine this special power is not definitely known. They apparently have a greater preponderance of the luminiferous ether than is common, but then what *is* luminiferous ether? Many psychics hold their vocation reverently. Many hold it commercially only, and, as we all know, some are entirely sincere and truthful, and some are not. Many people draw a strict line of demarcation between the professional and the nonprofessional medium, declaring that they have no faith in the former. Does not this seem unreasonable? If a medium devotes his (or her) time entirely to this calling, why should it not be remunerated as is the calling of the ministry? As the world goes, it must be. The medium must pay his bills like other people; and if he devotes himself to this calling he is entitled to just payment, nor does this any more invalidate his spiritual usefulness than the salary of a clergyman invalidates his usefulness to his parish. As a matter of fact, the professional medium is apt to be more unerring as a transmitter of messages than is the unprofessional. For mediumship, like all other vocations that have to do with either the material or the immaterial world, grows stronger by definite practice.

It is precisely the same with the vocation of the poet. Mrs. Browning used constantly to urge upon her husband, during all the years of their married life, — that wonderful idyl of fifteen perfect years, — the desirability of going to his study immediately after breakfast with the definite intention of writing poetry. To her it was a calling, a vocation as well as a consecration. "I never mistook pleasure for the final cause of poetry," she said; "nor leisure for the hour of the poet." Every worker in any line whatsoever, in poetry and romance as well as in the less inspirational order of literary work; in spiritual seeking and in prayer, as well as in official and mechanical and industrial pursuits, knows the untold magic of regular hours and a definite purpose. "No work that is worth doing," said one of the greatest of men, "can be thrust into the holes and corners of life." Mr. Browning was not, however, temperamentally amenable to Mrs. Browning's suggestions. He was variously gifted, and during all his earlier life music and sculpture attracted him almost as strongly as poetry. The artist suffers when he is the victim of over-possession. His efforts in any one direction are neutralized, if not paralyzed, by counter-attractions. A body placed at the center of the earth would be equally attracted in all directions and would therefore remain motionless. The too numerous attractions are equally disastrous to specific achievement. Whether, after Mrs. Browning's withdrawal from the visible world, she was able to influence her husband more potently must remain an unanswered question; but, at all events, it was after she had vanished that he entered upon regular morning hours for work, and that he produced his greatest poem, "The Ring and the Book." It was in the spring of 1860, more than a year before her death, that he had chanced upon "the old yellow book", when strolling through the piazza of San Lorenzo, on a market day; but it was four years later before he had transmuted the tragedy of the Franceschini into his immortal work.

The professional psychic who brings to the vocation the added potency of attention focused, as it were, at regular hours, is apt to be more unerring as a transmitter than one who only exercises the gift at irregular intervals. But surveying the entire field of mediumship from this present vantage point of time, one could hardly escape the conclusion that mediumship has been a phase, a temporary bridge, a lamp in the darkness; but that now the time has come, or is rapidly approaching, when it is no longer needed. Nothing can be more unsatisfactory, as a rule, than the *séance*. It has served a great purpose; but its best use was to lead to its disuse. It has served to establish the indisputable fact that communication between the two states of life is possible; the complaint that it has never given any communication of value is unfounded; it has given, first and last, during the seventy years of modern spiritualism, a proportion of communications of significance; and it has given a very great number of communications that have established the identity of the communicator, although nothing of much importance was said. The establishment of the truth that communication is possible is the all-important purpose it has served. After that, the messages, however interesting or comforting, are yet negligible compared with the fact that messages are possible at all.

Now that the purpose is served, — then what? The next step is for each to so develop his own spiritual faculties that he may be in telepathic response to his friends in the ethereal realm. The higher being, the spiritual self, the real self in every person can be awakened. But this awakening can only be accomplished by the individual for himself. He must generate the force that will unlock currents of energy hitherto unsuspected. He must generate the force that will set free a higher range of faculties. It is the liberation of this force that is known in religious experiences as conversion. It is a very real fact of life. It may easily be the supreme fact and the transcendent experience of life, an epoch, that ushers one into a new world. It was an experience of this order that Edward Everett Hale thus describes:

"I began by seeking during the day one hour of perfect solitude. As the weeks went by, I began to be conscious of a curious change in myself which I did not and do not explain. My pleasure in the many interests that made up my life began to diminish and become dull. Instead of desiring to finish the duties to turn to the pleasures, I found that the so-called pleasures had little interest. Various things that had filled my mind lost attraction. I felt no lack in life, however. I believe I was conscious of a greater interest."

The poets have always testified to the reality

of the spiritual realm that encircles humanity. This testimony has not impressed the general reader with its true significance. It has been relegated to the atmosphere of imaginative romance. Yet to the poet (the very perception and experience, indeed, that determines him as a poet), the reality of the interblending worlds is invariably recognized. No writer of verse who has not this recognition and conviction has poetic immortality. His songs may have a season of æsthetic recognition, but they hold no enduring spell over the minds of men. All poets who have won universal recognition are poets who intuitively and inevitably affirm in their work the reality of the spiritual life. One does not need to offer in evidence any list of names to support this assertion. No poet has expressed his perception of the ethereal realm as interpenetrated with our own more clearly than has Lowell in the lines:

"We see but half the causes of our deeds
  Seeking them wholly in the outer life,
  And heedless of the encircling spirit world
  Which, though unseen, is felt, and sows in us
  All germs of pure and world-wide purposes."

Although in previous writings I have (perhaps more than once) quoted these lines, they are instanced here as more perfectly embodying the ideal of the twofold life possible to each and all than almost any other passage from any poet. It is in this expression that one may find the true meaning of the term "spiritualism." It is not in phenomena, not in tables rising in the air, not in raps, nor in bells rung in the air, nor lights seen that proceed from no normal source, — it is in none of these things that the faith is to be sought. There is a world of legerdemain, of necromancy; there is also a world of physical phenomena, of which such intelligent experiments and investigations as those of Doctor Crawford offer legitimate interest; but it is not in these phenomena that spiritual aid will be found. Spiritual things must be *spiritually* discerned. The fact that forces in the ethereal world can (and do) transcend physical laws and thus reveal the existence of a higher range of laws in physics than those we yet know, — this fact has no more to do with spirituality of life and with communion with friends in the unseen than has any chemical experiment that might be made, however interesting in itself.

It is the quality of the communion enjoyed that is important. It is an interesting scientific fact that a man in New York may speak to another in San Francisco; but this speaking is not to be mistaken for the leisurely conversation with its mutual thought and sympathies. The analogy holds true in the contrast between the receiving of a message through mediumistic aid and the prolonged telepathic communion possible to those attuned to the same key of vibration.

Life in the ethereal is in perfectly natural relation to the life in the physical world. During this past seventy years of modern psychic phenomena much definite information has been given as to the conditions under which life in the ethereal moves on. That there is no such contrast to the conditions here as has been supposed seems sufficiently attested by the mass of evidence that many who have passed out do not realize the transition.

All nature has two aspects, the material and the ethereal, which as strictly correspond as do an object and its reflection in a mirror. To adjust the mind to the realization of this natural condition, to speak to those in the unseen as one would speak to a friend in the same room, is to enter on an order of communication that is full of solace and joy. Where is this ethereal world? It is in your room, your home, your grounds; it is in the streets of the city; it is in the woods and the mountains; it is on the sea; it is everywhere because the ethereal and the physical worlds interpenetrate.

V

HOW TO DEVELOP SPIRITUAL RECOGNITION

"My spirit to yours, dear brother;
I do not sound your name, but I understand you."
— WALT WHITMAN.

"When two clasp hands and part, they go toward the
future meeting;
For the path of life is a circle; be sure they shall meet
again." — ELSA BARKER.

IN "Aurora Leigh" Mrs. Browning has something to say of the value of keeping up open paths between the seen and the unseen. The power of any individual life is indefinitely multiplied by the aid of clear and well-defined views of its relations to the ethereal realm and its possible extensions into the unseen. These extensions are practically unlimited. Just as one may have all the air he can breathe, without money and without price, so may he draw from the ethereal realm all the potency he can appropriate. The only limitation is within himself. There is none on the other side. He may draw on these forces for health; for successful achievement; for power to help others; for knowledge; for spiritual vitality. And he will find that the promise, "To him that hath shall be given" is particularly fulfilled in this relation. As one draws from this infinite reservoir of power he learns how to draw more; as he assimilates and appropriates these energies, and applies them to specific purposes, he learns how to assimilate and appropriate still greater potencies. Saint Paul, enjoining that men "might be filled with all the fulness of God", adds this impressive statement:

"Now unto him that is able to do exceeding abundantly above all that we ask or think, according to the power that worketh in us."

The last clause indicates the condition of receiving abundantly. It is "according to the power that worketh in us." And this power is faith. Faith creates the condition by means of which the divine aid can come. Faith is a creative energy. It is a great fallacy to suppose that faith is a merely passive mental state in which one idly waits for some miracle to happen to him. On the contrary, it is a condition of the most intense form of energy. The Catholic expression of an *act* of faith is significant. It *is* an act; it is doing something, when one has faith. It is a process of spiritual creation. God is able to do, "exceeding abundantly" all we ask, *if* we do our own part. But, as the apostle so clearly portrays, this divine aid is according to the power that worketh in us.

Emerson suggests the ideal condition of living when he says, "Every touch should thrill." One must so order his life, physically as well as spiritually, to the end of keeping in sensitive response to the vibratory influences of the ethereal realm. The philosophy of fasting was to bring the physical nature into this more sensitive and subtle response. While man inhabits his physical body its condition greatly limits, or promotes, the power of the higher influences. It may almost exclude them from his perception. The bodily condition renders the man more or less impenetrable or responsive. So it comes to this: that if any physical habit or self-indulgence tends to more entirely imprison the spiritual self, then it is wrong simply because of that effect.

Phillips Brooks was once asked how certain things seeming innocent enough, not to say quite negligible, in and of themselves could be wrong? The reply of Bishop Brooks was to the effect that if things not wrong in themselves yet kept us from better things, to that extent, then, we must class them as wrong.

The teachings of Theosophy regarding the nature of the physical body and its relation to the ethereal body have for their purpose the presentation of knowledge and aid to the establishment of open channels for the divine energy to reinforce and recharge the human energy. The physical body is very plastic, and its matter can be modified constantly by the force of the will. All hygienic science has for its objects and results the more complete domination of the physical mechanism by the power of the spirit. College athletics are not an end in themselves; but athletic culture gives to the youth a power of control over this physical instrument that is of untold use to him. Theosophy contemplates man as a dense body, a vital body, a desire body, and, with other intervening states, to at last achieve the spiritual body. During the evolutionary progress of the spirit, the outer bodies, in successive relays, become finer and still finer as the spirit exercises upon them its increasing control. Spiritual potencies are constantly transmuted to dynámic energy.

The standpoint of the Christian Fathers was that while it was hard to fight poverty and hunger, yet from the standpoint of the soul's progress these were far preferable and far more favorable than luxury. It is left for the more advanced civilization to realize that comfort and ease may be so held as to minister to the higher life; to facilitate achievement; and that, as Emerson tersely says: "A cushion is good if you do not use it to go to sleep." We have learned that there is nothing inherently immoral in wealth, or in the larger privileges and opportunities that it opens; it is the use we make of these opportunities and privileges that determine the matter. Thought force is the most intensely creative of all potencies. Create in thought; to realize this creation in the outer and objective life, is the invariable process. The unmeasured potency of prayer, as the means of uniting man with his higher self and uniting him with the divine life, is a potency that exceeds all definition or human comprehension. It is the power that leads man on from glory to glory. It is this power that develops spiritual recognition.

Desire, alone, effects nothing. Will, purpose, must be brought to bear. To bring the physical mechanism into complete harmony with the controlling thought; to so refine and dominate it that it will serve as the most delicate and flexible and sensitive instrument to transmute plan and purpose, is the object of both hygienic science and moral law. When one comes to study the various occult sects and cults, the Rosicrucian, the Theosophical, and others, one finds the basis of each and all, so far as discipline is concerned, to be that of making the body serve as the perfect instrument of the spirit. That is the use for which it is designed, and its temporary nature is simply because that when the spiritual man withdraws from the physical world he has no

further need of the instrument that related him
to that world.

The spiritual forces play a far larger part in
this unexplored universe in which we find ourselves than we recognize. We are, indeed,
"heedless of the encircling spirit-world", and it is
as we apprehend more clearly its part in daily
life that we become more efficient. Science has
revealed to how limited an extent we see the world
in which we are placed. The telescope and
field-glass reveal a wider range on the one side;
the microscope reveals a wider range on the
other side. Now there is no inherent im-
probability in the speculative conception that
those who have died are still dwelling to a greater
or lesser extent in the same space in which we
find ourselves. That we do not see them is no
argument against their possible presence. The
eye only registers within its own degree of
vibration. The ethereal body, as we have seen,
is invisible, that is to say, unregistered by the
physical eye, because its rate of vibration is
beyond the range of that registration. But that
their sight includes us, in part, or at certain times,

at least, seems to be established. This would account for many warnings of danger; for many suggestions that find their way, by one means or another, to those here. Whether this power of cognizance is associated with actual presence in the sense in which we understand that; whether it is telepathic and may proceed from any point in space, is problematic. But the result on this side is much that of the close presence as we should understand it here. How, then, shall we develop our recognition of that cognizance and our own ability to respond to it?

There are possibilities of resource in the ether beyond man's comprehension. The ethereal currents that make possible wireless telegraphy were as much in the atmosphere when Columbus discovered America as they were when they were discovered by scientists four centuries later. Who may venture to predict the nature of future discoveries in nature? The spiritual man exists independently of his physical body. He is capable, even before death, of partial detachments from it. The spiritual man has faculties undreamed of in the present. He possesses a power, latent to a great degree, to attract new forces, to alter conditions, to act upon existing phases of the outer life. To this end Faith seems to be the key. Doubt disperses and dispels and destroys power. Faith fosters the power until it grows as the mustard seed and becomes a creative force. Now this power to act upon events and to bring one's self into harmonious receptivity to the divine currents may be largely assisted by friends in the unseen. Thus may those in the two conditions bring to bear the best energies of both states of life. It is not improbable that the youth who have passed from the front into the next phase of life are still contributing aid beyond that which was possible for them to give here. Jamblichus, who died about 333 A.D., said, even in that far-away time;

"If the soul rises to the gods she becomes godlike, and able to know the above and below; she then obtains the power to heal diseases, to make useful inventions, to institute wise laws. Man's intuition is the result of the connection existing between his soul and the Divine Spirit;

the stronger this union grows, the greater will be his intuition or spiritual knowledge. . . . If the mind of man is illumined by the Divine Light, the ethereal vehicle of his soul becomes filled with light and is shining."

Not only from the early Christian centuries, but from periods long antedating the appearance of Jesus on earth, similar testimony comes. The perception of spiritual truth advances as man advances in development. The twentieth century should give us a larger view; nor is it venturing too much to believe that this larger view already manifests itself in the world. The magnitude of the War, its unprecedented depths of tragedy, are bringing us face to face with spiritual realities. Consciousness is extending itself to hitherto unexplored regions. Man is learning to send his soul through the invisible. In proportion to this extension of consciousness is man's approach to larger truth. The larger view of truth promotes greater effectiveness in all the affairs of life. There is no limit to the radius to which consciousness can extend itself. Spiritual advancement is as recognizable a fact as advancement in electrical science. And as consciousness extends itself toward the Infinite Consciousness, man grows more capable of co-operating with the divine purposes, and it is thus, in the language of the Bible, that he may "walk with God." Archdeacon Wilberforce made the striking assertion that "The human soul is a dynamo, generating spiritual electricity from a magnetic field as vast as the whole universe."

Should we not, then, be able to penetrate with intelligence and accuracy to some degree beyond the confines of the physical world? May we not enter upon cosmic truth? May we not discover that the universe of all intellectual and spiritual life is one; that in this universe those in the physical body and those who have withdrawn from it are all dwelling together? Love itself unites closer bonds in this realization.

> "Regret is dead, but love is more
>   Than in the summers that are flown,
>   For I myself with these have grown
> To something higher than before."

Again, we find Tennyson saying:

"Known and unknown; human, divine;
  Sweet human hand and lips and eye;
  Dear heavenly friend that cannot die,
Mine, mine, for ever, ever mine."

Spiritual recognition, therefore, is attained by rising into the realm of the spiritual order. "Why do we make no greater advances?" questioned Mrs. Browning regarding communication with those beyond. "Why are our communications chiefly trivial? Why, but because we ourselves are trivial. Why, but because we do not bring serious souls and concentrated attention and holy aspirations to the spirits who are waiting for such things? . . . What comes from God has life in it, and certainly from the growth of all living things, spiritual thought cannot be the exception."

Poet and seer unite with prophet and apostle in the conviction that the exaltation of our own life is the condition of the recognition of spiritual realities. Communication, spirit to spirit, should be one of the channels of religious progress.

There is a wide contrast between the simple truth of spiritual companionship and the mysteries of occult phenomena. People have grown bewildered, if not repelled, by the rehearsals of the *séance.* To identify the beauty and naturalness of intercommunion with a mass of objective phenomena, — with raps, with alleged materializations, with the ouija board, with crystal-gazing and other forms, — is a confusion that strikes dismay to the minds of many. These forms of manifestations from the unseen are all genuinely used (whatever may be occasional fraud or imitation); but in the higher and larger aspect of spirituality of life they become negligible.

The danger in all this objective phenomena is that of inconsequential communication, as there might be were the doors of one's home freely opened to any miscellaneous passing crowd. While there are not wanting authentic instances of communication through a psychic that is of both comfort and value, it is still true that the better way is to learn to receive the thought, the expression, through one's own spiritual faculties. Archdeacon Wilberforce, who was left in desolation and loneliness by the death of his lovely wife although continually conscious of her uplifting sympathy and presence, sought a definite communication through the mediumship of a very remarkable woman, Mrs. Etta Wriedt, who had gone from her home in Detroit to London at the invitation of Mr. Stead.

The three *séances* that the Archdeacon had with Mrs. Wriedt were very remarkable. He was a trained observer, but he was also a man of the most delicate and unerring spiritual perception. Many sceptics and doubters who believe themselves critical are, instead, dense. They are too unawakened to the spiritual side of life to recognize truth even when presented. The Archdeacon was not a man to be easily deceived, nor, on the other hand, one to fail in recognition of any genuine communication. Through Mrs. Wriedt's powers the audible voice is heard; "and," said the Archdeacon to the writer of this book, "if ever I heard my Charlotte's voice, if ever I talked with my wife, I did on these occasions." Had it been merely the voice alone, however unaccounted for save on the theory that Mrs.

Wilberforce was speaking, there might be room
for discussion if not for well-founded doubt;
but the contents of those conversations included
matters known only to the husband and wife
themselves and were of a nature to entirely refute
any possible theory save that Mrs. Wilberforce
was speaking.   Then, too, the Archdeacon related,
even quite aside from the subject matter, there
were turns of expression; allusions; a thousand
subtle things, incommunicable as "evidential"
matter at the stern and rigorous bar of the Society for Psychical Research, but inevitably the
strongest and most unmistakable proof of identity
to the Archdeacon.   It would not be right nor
just, when Mrs. Wriedt, Mrs. Soule, and others
of a high order, such as Mrs. Piper of Boston,
whose fame as a transmitter of messages from
the beyond is world-wide; who is the honored
friend of Sir Oliver Lodge — it would not be just
when these exceptional psychics, and others, too,
that might well be named, are proven so genuine, ,
to fail in appreciation of this order of service.
Yet it may be (and, for one, I believe it is) the
ideal for each individual to so develop his spiritual

faculties that he may be in direct and personal touch with the unseen. This achievement is already much in evidence, and it will become more and more universal.

Mrs. Livermore (and a saner or more poised woman than Mary A. Livermore could hardly be known) used to say, after the passing of her husband, the Reverend Doctor Daniel Parker Livermore, that every morning, after finishing her correspondence and meeting other immediate demands, she could call her husband and pursue an intelligible conversation with him, his part in it being instantaneously impressed upon her mind as naturally as if it had fallen audibly upon her ear. The time is perhaps not very far distant when Mrs. Livermore's experience will cease to be exceptional.

No means of developing spiritual recognition, aside from prayer, always the most intense power in life, can be so helpful as that of taking a certain time alone each day to lift up the heart and thought and to give one's self to the higher currents of the diviner atmosphere. This practice sets free the higher powers.

But it is with life, the quality of daily life, that we are most concerned. "The field is the world." The test is in the average daily contact, in work, in social life, in incidental meeting and encounter. The test of spirituality of life is in the homely virtues of honesty, truth, justice; it is in the unconscious influence exerted; it is in the effort to make one's self a link to carry forward hope and happiness. The hour of uplift and meditation; of opening the mind to all nobler calls; the hours even for prayer, are still means to an end, not an end in themselves, and that end is in diviner living.

It may be confidently held that

> " . . . . Life is ever lord of death,
> And Love can never lose its own."

Where there is a spiritual bond there can be no separation. It is indissoluble for time and for eternity. We shall follow those who precede us into the ethereal world. What does Emerson say?

> " 'Tis not within the power of fate
> The fate-conjoined to separate."

Love is of the immortal life, and over it neither time nor change nor death has power. "Love is watchful, and, sleeping, slumbereth not. Though weary, it is not tired; though pressed, it is not straitened; though alarmed, it is not confounded; but as a lively flame and burning torch it forces its way upwards and securely passes through all.

"Love feels no burden, thinks nothing of trouble, attempts what is above its strength, pleads no excuse of impossibility; for it thinks all things lawful for itself, and all things possible. It is therefore able to undertake all things and warrants them to take effect, when he who does not love would faint and lie down.

"He that loveth, flieth, runneth, and rejoiceth; he is free, and cannot be held in. He giveth all for all, and hath all in all; because he resteth in One highest above all things, from Whom all that is good flows and proceeds. Love is active, sincere, affectionate, pleasant, courageous, faithful, and never seeking itself.

"If any man love he knoweth what is the cry of this voice."

Love is an inner and all-pervading and a trans-forming energy. It can achieve the impossible. It can endure the unendurable. It can create life anew from ruins. "Sorrow is a condition of time, but joy is the condition of eternity," and Love discerns the eternities. The mission of Jesus was to bring life and immortality to light; "I am come that ye might have life and have it more abundantly;" for life here on earth lived divinely is far more abundant than the ordinary human life; and immortality teaches that death has no terror, being merely the process of tran-sition into the fuller life and joy beyond. The life beyond this transition bears the same relation to our present life that youth may bear to infancy and early childhood; that mature manhood may bear to youth. The evolutionary progress is continuous, gradual, unbroken. Who can dis-cern any crisis day in the development of the infant to the man? Yet the transition goes on before the eye. The normal and orderly devel-opment of life includes mutual companionship between the two states. All phases of progress here imply somewhat of conquest over the ethereal conditions. Sir Oliver Lodge, speaking of this matter has said:

"If there is any object worthy the patient and continued attention of humanity, it is surely those great and pressing problems of whence, what, and whither that have occupied the attention of prophet and philosopher since time was. The discovery of a new star, or of a marking on Mars, or of a new element, or of a new extinct animal or plant, is interesting; surely the discovery of a new human faculty is interesting, too. The discovery of telepathy has laid the way open to the discovery of much more. Our aim is nothing less than the investigation and better comprehension of human faculty, human personality, and human destiny."

Telepathy is simply the spirit language.

"Star to star vibrates light; can soul to soul
Strike through a finer element than its own?"

Soul to soul can, and does, strike through this finer element. The tragedy of the War, the stupendous nature of the international conflict that began with the August of 1914 and which closed in the early November of 1918, is revealing more impressively than it was ever revealed before the truth of communion unbroken by death. It is a truth that will revolutionize all the philosophies in the world and will largely modify, if not transform, the systems of education. For children will be taught the true nature of our relations to the unseen. Death will no longer be regarded as a mysterious terror. Through this philosophy the spirit of man will have been lightened and exalted and enabled to increase in spiritual energy.

VI

DAILY LIFE TRANSFORMED BY SPIRITUAL VISION

"A Divine light strikes upon me, penetrating through this wherein I embosom me; the virtue of which, conjoined with my vision, lifts me above myself so far that I see the Supreme Essence from which it emanates. Thence comes the joy wherewith I flame, because to my vision, in proportion as it is clear, I match the clearness of my flame. . . . O joy! O ineffable gladness! O life entire of joy and peace! O riches secure, without longing! . . . Behold now the height and breadth of the Eternal Goodness!"

— DANTE: *il Paradiso.*
(From the prose translation by Charles Eliot Norton.)

Thy testimonies are very sure: holiness becometh thine house, O Lord, for ever. — PSALMS: 93: 6.

THE Beautiful Days are approaching. Every hour brings them nearer. For in proportion to the distance that these Beautiful Days receded and their experience seemed to fade beyond possibility of recovery, — in just this proportion they are advancing to us and we are approaching to them.

153

"For the path of life is a circle."

We are about to enter on the new order. Human life has been incalculably elevated and ennobled by tragedy, sacrifice, suffering. Let us not only keep faith for it, but keep faith with it. For faith is divinely creative and is the condition of realizing that in which it believes. Let us keep hope; let us approach the new order with courage. With Lowell one may say:

"I have no fear
Of what is called for by the instinct of mankind."

What is this unknown future into which man is advancing? It is deliverance and salvation. For two thousand years the Christian world has prayed to be delivered from evil. The gradual deliverance, the larger elimination of the evils of life are at hand. We are on the threshold of a world rich in deeper experiences; glorified with higher hope and purpose. New stores of cosmic energy shall be unlocked. Man's intellectual power increases in proportion as he advances into this ethereal world. The history of the progress of spiritual brotherhood is the history of social evolution.

Material substances have been regarded as the substantial ones out of which to fashion the enduring monuments and structures of earth. But beyond these is the still more enduring and more potent substance of Thought.

Fundamentally, all things are made by thought and will. To create in brick and mortar is a slow process; to create in thought is instantaneous. This higher creative power is about to be made so applicable to the conditions of life on earth as to produce a marvelous change in all industries. Had it been prophesied in the early years of the nineteenth century that the human voice would be heard from New York to San Francisco, from Washington to Hawaii; that messages between Europe and the States would flash under the ocean; that messages sent through the air on a ray of the ether without visible mechanism, would be transmitted around the entire globe, who would have believed such a forecasting? Yet within half a century all these things have become common knowledge

and common practice. Man is on the threshold
of changes still more extraordinary because he
is about to enter into the realm of higher law.

The resources of the ethereal realm are infinite.
In the ethereal energy lies all constructive power;
all possibilities of instantaneous communication; all possibilities of a new order of transit.
The spiritualization of matter is the next onward
step in civilization. Henri Bergson perceives
this truth. He argues that life should be free,
spontaneous, that while it is now clogged and
hampered by matter, its free creative activity
is the ultimate reality. Monsieur Bergson has
also offered a speculative theory that is, at least,
one of curious interest. It is that consciousness,
which he regards as one great unity, pours itself
with resistless force through separate individualities; that matter, or the soul, being immersed
in and clogged with matter, is what keeps back
the rush of life; that man has but to remove the
obstacle and more consciousness rushes through.
"Organize individuality a little, and a little life
will pass through. Organize it still more highly,
and the more consciousness, the more life. Or-

ganize it elaborately, and still more life will come through." It is a common experience to perceive that some men are more alive than others; do we find the explanation in the theory of Henri Bergson?

Arthur James Balfour has asked the question: "Is the flood of life really beating against matter till it forces an entry through the narrow slit of undifferentiated protoplasm?" And he also questions as to whether it is possible for philosophy to establish the reality of this theory. "Bergson's '*Évolution creatrice*' is not merely a philosophic treatment," continues Mr. Balfour; "it has all the charms and all the audacities of a work of art, and as such defies adequate reproduction. Yet let no man regard it is an unsubstantial vision. It mingles minute scientific statement with the boldest metaphysical speculation. His philosophy never wearies of an appeal to concrete science."

Mr.. Balfour points out that Professor Hertz demonstrated experimentally the identity of light and of certain electro-magnetic phenomena. Now light consists of undulations of the luminiferous ether. Electro-magnetic waves are also found to be undulations of this same ether, differing from the undulations of light only in length. Mr. Balfour then calls attention to this fact: that if man had a sense by means of which he could perceive the long undulations in the same way that he perceives the short ones, this would be a new sense and open to him a new world.

Are we, then, on the very threshold of this new world? Will not this higher life begin to impose itself on the ordinary life? "The electric theory," says an English authority, "carries us into a new region altogether; it analyzes matter into something that is not matter at all, postulating nomads as units of electricity." Theosophy states an illuminating truth in the following affirmation;

"The invisible worlds interpenetrate the visible, the crowds of intelligent beings throng round us on every side. Some of these are accessible to human requests and others are amenable to the human will. Christianity recognizes the existence of the higher classes of Intelligences under the general name of angels, and teaches that they are 'ministering spirits'; but what is their ministry, what the nature of their work, what their relationship to human beings? — all that was part of the instruction given in the Lesser Mysteries, as the actual communication with them was enjoyed in the Greater, but in modern days these truths have sunk into the background."

Professor Tyndall found that the luminiferous ether is so attenuated and elastic that it can convey vibrations of light at a rate of some two hundred thousand miles a second. If man had the faculties developed to enter into relations with such an atmosphere as this, his environment would be completely transformed. Life would then be in the higher etheric vibrations of spiritual substance. The microphone demonstrates the actual presence in the atmosphere of innumerable waves of sound of which the physical ear takes no cognizance. In this realm of finer and higher vibrations, too subtle to be registered by the ear or the eye, may not spirit voices sound? May not the ethereal bodies live and move? Such philosophers as Stewart and as Tait postulate the existence of an unseen universe, with the strong presumption that it is full of life and intelligence, that it is infinitely higher in its degree of intelligence than the universe we know, as it is infinitely more potent in force. Only beings of a higher organization could exist in this environment. Stewart and Tait contend that we must resort to this subtle universe for an explanation of the forces that carry on the universe in which we live. To a wonderful extent, here and now, the regeneration of the body can be effected by the renewal of the mind, according to the literal counsel of Saint Paul. The secret of this renewal is in being able to exercise the power to bring currents of consciousness into connection with the vital cells of the body. It is entirely possible, if one may learn the way, to maintain the physical mechanism in a state of unbroken health, harmony, and energy. It depends upon spiritual initiative.

The new order of human experience thus faintly outlined and fragmentarily suggested is that which lies just before humanity at the present time. "It doth not yet appear what we shall be." But apparently it is a preliminary necessity to sweep away old conditions. Industrial and social problems will be reinterpreted and readjusted. May it not be that the Power which makes for righteousness employed even the tragic means of this recent conflict in order to carry humanity to a higher plane? Sacrificing the kingdoms of the material and the temporal, man advances into the kingdom of the spiritual.

The vast numbers of young men who so suddenly passed over from the front, carrying with them such devotion and love, are bringing the life beyond into familiar comprehension. They entered there in the spirit expressed by Dante:

"O splendor of God, by means of which I saw the high triumph of the true kingdom, give me power to tell how I saw it!"

They return to assure those who follow them in the unbroken consecration of love that the world they enter is as natural as the one they leave, and that there is no break in the unity of life. They go in joy and triumph. To his

mother, just before death, a young soldier wrote:
"When I enlisted I knew such a day as this
might come, but I do not regret it. I am happy
in the thought that I can make my gift complete.
Will you not try to be glad and thankful with
me?"

One communication from a soldier was given
by automatic writing to Mr. T. N. Brocas, of
Auckland, Australia, and was published by the
recipient in "The Harbinger of Light", a journal
in Melbourne. The soldier wrote:

"I am trying to give you all a true and direct
account of what has happened to me on this side
of life — that is to say, since I left the earth plane
on being killed at the Dardanelles by a Turkish
bullet, as you have no doubt heard already.
After I sent those shawls to you I was for some
time in Egypt, but directly after sending those
last two postcards I, with many others, was
sent to the Dardanelles to fight the Turks. . . .
I commenced to run, with my bayonet ready at
the charge, when I felt a tremendous shock,
and then all seemed dark for a time, but how
long I don't know. Then I awoke to find myself

standing among strangers. Some seemed to be my own people and some seemed like the Turks.

"I turned to some of those near me and said, 'Where am I? How did I come here?' and 'Where is the fight? I cannot see or hear anything of it, or my companions.'

"They smiled, and one of them said, 'We are as strange as what you are, and don't know how we came here; but I suppose we have been ill and have been brought here while unconscious.'

"But directly after this a strong, active man came, quite suddenly, and said, turning to me and those near to me, 'Do you not yet realize that you are all dead?' and he smiled such a smile. I said, 'Dead! No! I am not dead! Indeed, I am very much alive, I can tell you; but I don't know how I came here. The last thing I can remember is charging at those deadly Turks, then I felt a shock and woke up here to find myself in a strange place.' I found that I was really dead. Well, that is to say, I had come over into the other side of death, into life, and I can tell you, dear friends, it is a life, and a greatly better life, than the old one, for there is no more death to fear and look forward to. Don't be afraid of death any more; the only sting of death is the temporary parting from those we love, but even that is softened to a great extent, to some at all events, for they are allowed to get in touch with their dear ones to some extent.

"I cannot tell you much, but I have met my mother, and she and I had so very happy a meeting; but we sorrowed over the fact that father would be grieving over my death. But, oh, it will not be so very long till we are all united.

"I must go, but I will come again later on, and will try to tell you more about our life over here, and do believe I am really trying to talk to you all."

A series of messages from a soldier to his mother, recently published in a small book,[1] offer an unusual example of fact and incident from the unseen. Before he went to the front the youth had been an enthusiastic experimenter

[1] "Thy Son Liveth." Boston. Little, Brown, and Company, 1918.

in wireless telegraphy. The apparatus was left in his room, and he had half laughingly said to his mother, before he went, that he would find a way to send her a message through it; this promise, however, having to do with his life "Somewhere in France" and not in the ethereal world. But it was from the latter that the first message came. His mother had gone to his room to read a letter from him which had just arrived, when suddenly the apparatus signalled "Attention." She sprang to the key, — she had before this learned the code, — and the message came, beginning:

"Mother, be game. I am alive and loving you. But my body is with thousands of other mothers' boys near Lens."

Transcribing this, the mother wrote:

"So the news that my son had been killed came to me from his own intelligence by the methods we had used together in our experiments in this very room. . . . I have no explanations or proofs other than those that are given here. *A man who was killed in battle and is yet alive, and able to communicate with the one closest to him in sympathy, must make his own arguments.* I have no knowledge of established psychic laws or limitations. But I know what I know."

Aside from the wish to communicate with his mother, the special desire of this young man was to establish the proof of survival after the loss of the body in order to comfort other mothers and other bereaved homes. This motive, in both the messages from many sources, and their being shared with the public by those who receive them, is felt in common by all. If one family thus receives comfort they feel it a duty, as Sir Oliver Lodge notes in "Raymond", to pass this knowledge on and share it with all who are prepared to consider it. One thing that is continually emphasized by those in the ethereal side is the sorrow caused them by the mourning of friends on this side. "Every tear tortures the dead" is one expression in a message. "Try and make this point plain to the families."

To all who have close ties in the beyond, one chief source of grief is the thought that one cannot do anything any more for those so loved. It is perhaps true that we miss far more the privilege of giving some form of loving service or manifestation than we do the receiving of such manifestations and precious tributes. One who loves finds his dearest joy in doing something for the one beloved. But we can do infinite and wonderful things for those who have passed into the ethereal. We can do far more for them than was ever possible when they were on earth. For it is far more important; it offers far more of joy to the recipient to sympathize with his thought, to companion him in spirit, than it did in this life to offer him material tokens. And this companionship of spirit is so rich in its satisfactions.

> "*Now* I can love thee truly,
>   For nothing comes between
> The senses and the spirit;
>   The Seen and the Unseen."

For the first time, in the sweet relations of affection, the closeness of the spiritual relation transcends all others; and, as Lowell expresses it in the stanza above, there are no longer obstacles to come between.

First of all, the beautiful offering we can make to them is not to sorrow and grieve in a way that shadows and impairs all their new interest and happiness. Realizing the spiritual presence and companionship, we can share these interests and happiness.

In a lyric embodying much of truth occur these stanzas:

"How can I cease to pray for thee? Somewhere
　　In God's great universe thou art to-day.
　Can He not reach thee with His tender care?
　　Can He not hear me when for thee I pray?

"What matters it to Him who holds within
　　The hollow of His hand all worlds, all space,
　That thou art done with earthly pain and sin?
　　Somewhere within His ken thou hast a place.

"Somewhere thou livest and hast need of Him;
　　Somewhere thy soul sees higher heights to
　　　climb,
　And somewhere still there may be valleys dim
　　That thou must pass to reach the hills sub-
　　　lime."

In these latter years we are exchanging a faith that includes much definite knowledge for the former faith that included no knowledge at all of the conditions of life beyond. Science penetrates into the nature of the ethereal realm; spiritual perceptions on this side and the great mass of messages from those beyond unite in establishing some very clear conceptions of both the nature of life and its environment for those beyond the visible. A death in the household tends to draw each member of it into the radiant atmosphere. There is the strange, sweet sense of a different order of companionship; there are thought and message and feeling that flash between in telepathic form of expression. Shall not one then so enter into the spiritual loveliness of the transition that he shall walk in joy in conscious sympathy with his friend? For this is the priceless gift he may make, the service he may still render.

There is undoubtedly a deeper significance than we have been accustomed to give to the assurance of Jesus when He said:

"If ye abide in me, and my words abide in

you, ye shall ask what ye will and it shall be done unto you."

The words are not a vague and mystic phrasing that mean nothing in particular when analyzed. Here is a definite promise: "Ye shall ask what ye will and it shall be done unto you." But this promise is conditioned; and the condition is something marvelous. For what is it to abide in Christ? It is something more than to follow Him; it is nothing less than *the complete identification of the human self with the divine.*

The question readily arises as to whether such complete spiritualization of life is possible to any man while on earth. Does not the very question itself suggest that this spiritualization of life is not a question of environment, nor one in any manner conditioned by the objective world, but that it is the problem of spiritual achievement; of more and more entering into the spirit of Him who had conquered all lower inclinations and had thus become at one with the divine? To the degree, then, to which man, now and here, can thus enter into and merge his whole being

in God, *to that degree,* and no more, may he receive
the fulfillment of the promise, "Ask what ye
will, and it shall be done unto you." This
promise is on that plane of life from which all
selfishness has been excluded. The exclusion
of selfish purposes does not necessarily mean the
exclusion of what we call material things. There
is nothing inherently wrong in a material object.
It depends upon the use that it serves. In the
physical world material objects are our signs and
symbols; what are food, clothing, shelter, the
first necessities of aid to the distressed, but material things? For they may be divinely used,
as Jesus Himself divinely used physical aid and
relief. The entire purpose of life, — life in the
sense of its extension into all the infinite eternities, — is to increasingly lay hold on the divine.
To conquer the tendencies that drag us down;
to conquer selfishness, self-indulgence, injustice;
to live on the plane where we take the good of
another to be our own; where we joyfully sacrifice the lower that we may rise to the higher. It
is not too much to say that these lessons are
impressively imaged before man by the awful

tragedy of the conflict of nations. Its lesson of
self-sacrifice; of the sacrifice of the lower life
to gain the higher, is as unmistakable as the
Handwriting on the Wall. In the individual
instances are revealed the universal spirit. One
youth, himself the descendant of a Revolutionary
hero, leaving his studies at Harvard, made his
way to France as cabin boy on a cattle boat and
gained his admission to the *École d'Aviation
Militaire*. He wrote his name as a hero in the
battles of the air. He destroyed many enemy
air-craft. Then, on a golden September day
in 1918, while patroling the American lines,
came the fatal shot, and his body was tenderly
laid in a field "golden with buttercups." What
had this youthful spirit not achieved of the
ꜱublimest order of life, of the absolute partaking
of the divine life? "This I say," were his words
when he left, "that if I die, I will die fighting."
And the mother, learning of his death, could say,
"And what could be more glorious than to die
fighting the enemy? It was a glorious death
my son had, to glide down to earth on territory
held by the American troops after he had done

his best and given his all. The mothers of the United States and in all the countries are doing what God did. He gave His only begotten Son that liberty might have life."

Of such greatness of spirit was the power created that carried on the War. Was it nothing for a nation to rise from a life of easy pleasure and leisurely pursuits to such sublimity of soul as this? This one example which can be contemplated only through eyes dim with tears, but also with heart and soul uplifted in gratitude to the Divine Father that such splendor of spiritual exaltation is possible, is only typical of the spirit of all this Flower of Youth, — these young Knights of the Holy Cross, who go forth in the consecration of utter sacrifice of self that Liberty may be enthroned and triumphant. It is he who loseth his life that shall find it. Is it not true that the ineffable blessedness of abiding in the Christ is entered upon by such greatness of soul? Are we, then, as a nation, beginning to realize the actual significance of many of the divine promises whose deeper meaning has never before been revealed to us? "If ye abide

in Me, and my words abide in you, ye shall
ask what ye will, and it shall be done unto you."
The young men of all the nations who have thus
triumphantly and joyfully given their lives that
the nations may live are thus entering on a spiritual heritage, incalculable in its power and
glory. With what marvelous beauty and intensity of energies do they find themselves
after the withdrawal from the physical body,
which has served its purpose and is discarded.
Imagination falters before the vision of this
resplendent life just beyond.

"And they need no candle, neither light of
sun, for the Lord God giveth them light."

VII

"HERE AM I, LORD; SEND ME"

"Also I heard the voice of the Lord, saying, Whom shall I send, and who will go for us? Then said I, Here am I; send me."   — ISAIAH: 6: 8.

IT is this voice, it is this response, that we hear abroad in the land. The heavens are illumined by flashes of Brahmic splendor. There are sacrifice, privation, and sorrow. There are glad renunciations; there is a choral spontaneity of response to the voice of the Lord, "Whom shall I send and who will go for us?" The Divine Life is manifesting itself anew through the uncounted thousands of the youth who respond, "Here am I; send me." The moral grandeur; the intellectual illumination; the new sense of Immortality, — the marvel and glory of these new conditions of life through which all humanity is rising to a higher spiritual plane, mark this period as a crisis in all the history of mankind. This is the age, not of denial

175

and darkness, — it is the age of transfiguration. It is the process of the spiritual regeneration of man. These are the appointed conditions by means of which his latent higher faculties are being aroused.

From this age onward he is to be a new creature. He in whom this divine light has not flashed forth in an awakening is still asleep in the spirit, and can no more bring his forces to bear than a sleeping man can guide or prosecute a given work. Man must become aware of his higher consciousness. An ancient writer counsels, "Throw away your imperfections and become perfect in God." If ever in human history the hour had arrived in which such counsel as this might be considered in its fullest significance, it is in the present. The conditions are unprecedented in all the annals of civilization. This War was a great spiritual conflict. All the possibilities of future civilizations are being weighed in the balance. The issues are so vast, so incredible, that it would be strange if their very magnitude did not blind our eyes. The call to arms was the call to spiritual energy.

The soul of man is to be liberated; to be freed from the bondage of the many inadvertent errors of which, in easy and prosperous times, we took little notice. The little vanities and vexations of life; the unconscious selfishness of self-indulgences; the personal extravagance in dress, in appointments; the compromise with lower standards, — all these must go. And when they have gone mankind has thrown off a burden and a material weight. It is not that the soul would renounce art, beauty, poetry, all the loveliness of life. But she would renounce somewhat of artificial standards and requirements with which she has been impeded.

> "Then why pause with indecision
> When bright angels in thy vision
> Beckon thee tō Fields Elysian?"

To "throw away imperfections and become perfect in God" does not sound like so impossible a counsel to consider, — even to aspire toward, — in 1919, as it would have appeared in 1914; for these five years have wrought a signal change in the spiritual outlook. Visions, ideals, are

in the air. Dreams of a more perfect humanity
haunt the heart. "God's kingdom must come
and it is our business to see that it comes,"
the great and good Edward Everett Hale used
often to say. He had the soul of the prophet.
The time has come sooner than he would have
dreamed, when the literal fulfillment of these
words must establish itself. · For the full freedom of the nations implies the freedom of the
individual soul. It is the appointed task for
this age to create a new heaven and a new earth.
Each individual must become a temple of the Holy
Spirit, manifesting this hitherto undreamed-
of power. Why, this is not the call to loss, to
privation, to poverty of life, or effort, or spirit.
It is not the call to renounce all the culture, the
charm of life, all that we have held as so desirable
and essential in the past. It is the call to such
richness as man has never known. It is the call
to exalt culture and beauty and the enchantments
of life to a nobler plane. Is it any wonder that
this young knighthood instinctively recognized
the Divine Voice that was abroad in the land,
and sprang with eager joy to respond to its bid-

ding? How the flaming lines of Emerson make themselves heard anew;

> "So nigh is grandeur to our dust,
>   So near is God to man,
> When Duty whispers low, *Thou must,*
>   The youth replies, *I can.*"

For the past quarter of a century the world has heard much, through the ethical teachings of India, as scattered broadcast by the itinerant Swamis who have lectured everywhere, of the great benefits of the practice of man's union with his higher self, known under the Indian term of Yoga. The man's union with his higher self is being effected in ways unforeseen. · It has become the practical necessity of the hour.

All these forces are leading to a restatement of the Christian religion. The exclusion in this restatement is merely negligible; the inclusion of larger truth, or of a more perfect interpretation of the truth, is one of importance. It is to include faith in immortality, not merely as a religious expression to which the layman attached only a vague meaning, but as a vital and clearly comprehended fact of life. There will be included a recognition of psychical truth. There will be included the comprehension of the nature of the change we call death and an unquestioned conviction of the unity of the individual life in the physical and the ethereal worlds. We shall grasp the fact that the withdrawal from the physical body has no more power to change the man himself, in any instant way, than has the substitution of one costume for another.

With that closer walk with God for which the soul of Cowper sighed and which this restatement of religion will enjoin, will be included that easy, natural recognition of the presence of friends who have passed beyond, that recognition and telepathic communion of companionship to which much allusion has been made in previous pages of this little volume. It is not strange that when this companionship and communion is presented under the aspects of weird and incomprehensible physical phenomena the religious man should turn from it as something that desecrates that which he holds sacred; but seen in its true light, as a component part of our own spiritual life, just as social companionships and the sweetness of friendships are a component part of our life in the visible world, then will it be estimated aright. Then will it be seen as a part of the spiritual atmosphere of life presented by Jesus, the Christ. Man will come to realize not only that there is no such thing as death, save as a name defining a change of conditions in the onward progress of conscious life, but that this change causes no separation. No one has formulated the new and more extended view of truth into a clearer presentation than has Epes Sargent in the following propositions:

"(1) Man is an organized duality, consisting of an organic spiritual form, evolved coincidently with and pervading his physical body, having corresponding organs and developments.

"(2) Death is the separation of this duality and effects no immediate change in the spirit, neither intellectually nor morally.

"(3) Progressive evolution of the moral and intellectual nature is the destiny of individuals; the knowledge, experience, and attainments of earth life form the basis of the spirit life."

Mr. Sargent, a poet, a thinker, an accomplished man of letters, was the editor of the Harpers' " Cyclopedia of British and American Poetry ", the most notable, finely selected, and complete poetic anthology that existed up to the time of its publication in 1880. Since then a new school of poetry has arisen, of which, at that time, Walt Whitman was almost the only herald. Under the date of April, 1886, Doctor Hiram Corson wrote to Walt Whitman, saying, "There are points upon which I have been long pondering — one, especially, that of language-shaping, and the tendency toward impassioned prose, which I feel will be the poetic form of the future, and of which I think your 'Leaves of Grass' is the most marked prophecy."

Mr. Sargent's death occurred just before this important Cyclopedia was published. In the announcement of the volume the Harpers characterize him as a man of complex nature, high aspirations, and one whose profound knowledge of literature, whose clear, acute, and discriminating judgment eminently fitted him for this work, the crowning work of his life. In his spirituality of nature, as distinguished from the merely formal and academic, Mr. Sargent had the keenest and most unerring poetic intuitions. With this he united a philosophic bent; and in the early days of manifestations from the unseen world he had given serious and discriminating study to the phenomena. He had become convinced of the truth of communication between the two conditions of life in the physical and in the ethereal. He felt the truth that was later to be so well expressed by Doctor Charles W. Eliot when he said:

"The religion of the future will not be gloomy, ascetic, or maledictory; it will deal, not chiefly with sorrow and death, but with joy and life." Religion becomes joyful and vital and replete with creative energy when the manifestations of the spiritual universe are recognized in their true relation to the physical world. To restrict human perception to that of the physical senses alone limits man's world as the deprivation of sight and hearing limit the world of the persons thus afflicted. It is in proportion as man exercises his spiritual faculties that his world is enlarged and made more significant, more intense in its energies, more enthralling in its interests.

It is difficult to conceive of a statement more reasonable, or one that could more entirely commend itself to the moral judgment of the individual, than these three propositions formulated by Mr. Epes Sargent. That there is an organic spiritual form that exists entirely independent of the physical body, but which uses the physical body as an instrument through which to function during the sojourn on earth, has been abundantly proven both by science and by psychic study. That the process we call death is merely the separation of the man from his temporary instrument of communication with the physical world is abundantly recognized. That the progressive evolution of the intellectual and moral nature is the unending experience is a presumption supported by all religions; by all systems of ethics; by the intuitive recognition of the soul. Jesus came to bring life and immortality to light; that is, to make clear this fundamental truth of the endless process of spiritual evolution.

They who understand realize that the intelligent comprehension of man as a spiritual being has no relation to the idle and meaningless assertions made by those who have no definite conceptions of the true nature of life. Mere physical phenomena, genuine or fraudulent as they may be, are not a factor in the matter. The investigations and conclusions are on another plane. The true comprehension of the spiritual nature of man has to do with conduct, which Matthew Arnold rightly defined as being three fourths of life. It is a man's conduct which is the unerring touchstone of his degree of spiritual advancement.

The nature and conditions of life in the ethereal are becoming still more real, to say nothing of far greater and more universal concern, by the multitude of homes bereaved by the War. Love follows these vast numbers of young soldiers who died at the front into the experiences that immediately awaited them, — the conditions upon which they immediately entered. Communications have been frequent. Many of these are so linked with personal remembrances of their

life here as to be amply evidential, even to
the vigilant psychic researcher. They speak of
these conditions with the utmost naturalness.
They confront aspects which they do not understand and about which they speculate much as
they would here in entering on a new environment. In one of these communications we find
the young man saying that after a period of
helping on the battlefield they were to leave for
another place. "We did not fly, or float," he
says. "We just marched at a rattling good
pace. The only strange thing about it was that
we did not mind such natural 'obstacles as forests or rivers, but went right along through them
or over them. . . . We passed through vil-
lages shelled and destroyed. There were human
bodies everywhere. From this point of view
there is no more in death than removal from one
house to another." The communicator speaks
of their conductor — one who had been longer
an inhabitant of the ethereal — as apparently
receiving instructions in a way that puzzled the
newcomer. "There were no messengers or mechanical means like telephones or wireless. But

it seems we acquire the ability to hear anything addressed to us, personally, through any amount of space. That is how you reach us. And what we are trying to do now is to have you hear us as well as we hear you." [1]

This suggestion will particularly appeal to those who understand. There is, all in all, an accumulation of testimony that those in the unseen can, and do, hear the spoken voice. Then the next thing that follows is that those on earth shall also hear the voice from the unseen realms, and distinguish the spoken words. Clairaudience is the power of hearing with the spiritual sense. The words fall upon the mind with all the reality of tone and inflection. Clairaudience thus differs from the telepathic method, by means of which the thought is flashed upon the mind, but without this sense of tone and inflection. The young soldier from whose communications the above extracts are taken also said:

"I get all your messages, mother. I can only answer a few questions. Partly because I am

[1] "Thy Son Liveth." Boston. Little, Brown, and Company, 1918.

not yet sure of many things here, and partly because there seems to be no means of communication concerning certain conditions. That is, when we get beyond the usual, we are beyond the common medium of language. The words we know are inadequate to express our revelations."

This suggests that telepathy is of a higher and more universal order as a means of communication than clairaudience. The latter is limited in its scope to language as we know it on earth; the former has the infinite possibilities of the infinite universe.

In this world we find the individual life greatly enlarged and its capacities multiplied by the acquirement of new languages. The classics, the romance languages, open to man new worlds of life and of literature. They enable their possessor to enter into many phases of life and thought otherwise impenetrable to him. Is it unreasonable to infer from this that the ability to easily converse with those in the next higher state of life would be a signal advance in evolutionary progress? Removed from the association of the phenomenal, the inconsequential (as the phenomenal is but too apt to be), it would simply be a factor in the general enlargement of intelligence; an increasing comprehension of the universe in which we live; and the cancel-ing of the former mystery (not to say the terror) of death. It would thus eliminate the one great-est sorrow of human life. We should come to understand the nature of the change and know that it did not involve the separation of entire silence. It would be of incalculable intellectual benefit as well as consolatory. It would be far more; even that of the more intimate compre-hension of the Divine Wisdom.

"These things I have spoken unto you that in me ye might have peace," said Jesus; the words conveying the assurance that increased comprehension of the unseen life gave to man increased peace of mind and freedom from anx-iety. "In the world ye shall have tribulation," He added; "but be of good cheer; I have over-come the world." To have tribulation "in the world" does not mean that tribulation has geo-graphical assignments and is a factor in one realm inherently, and not in another. Tribulation is a condition of imperfect and defective spiritual life. The more completely man may unite his spirit with the divine order, the less his tribulation. He may endure privation, disaster,·but shall we not learn to distinguish between these and tribulation, which is the result of . mingled ignorance and selfishness. One may be hungry, or cold, or limited in a thousand ways of discomfort and inconvenience without being at all selfish or ignorant. He may so discriminate between temporary discomfort and onward progress as to enable him to patiently endure and vigilantly strive. To "endure as seeing the invisible" is of profound significance. It is the condition of faith that sees beyond the temporary, and faith is the creative power by which the immediate and temporary can be transmuted into the noble and the satisfactory. The "world" in which tribulation is a factor is a condition of spirit. Jesus overcame that lower condition; man may overcome that lower condition. When he rises into the larger spiritual life he has overcome tribulation. Rising into this larger spiritual life; feeling one's self a part of it, the sorrow for the dead, the grief and loneliness incident to the change, are transmuted to a new sense of the beauty and the joy of the new relations that have been established.

"Let not your heart be troubled," urges the Divine Teacher; reminding us that we already believe in God, and enjoining that we also believe in Him. For it is He; it is His personal experience and assurance that reveal to us the true nature of death. He demonstrated that this change had no power over the immortal being. "Now," He says, "ye have sorrow." That is, while uncomprehending of the nature of the great adventure, while still ignorant of its entire significance, "ye have sorrow"; then comes the assurance, "But I will see you again, and your heart shall rejoice, and your joy no man taketh from you." The words are as vital to-day as they were two thousand years ago. They appeal to us with the deeper meaning because man has advanced to a true comprehension of all that they mean. They lift us into the Blessed Assur-

ance; they open to us the celestial gates, even
the Gates of New Life.

Is it, then, that the final magnitude of the
result of the War shall be, — not only the establishment of more just industrial and economic
conditions; not only the promotion of temperance and the downfall of intemperance and
the evils in its train; not only the bringing about
of needed reforms and the promotion of better
social conditions; not only a renaissance of Art
and Literature, enriched and ennobled by all
the deepening of life in the world's tragedy; but
shall the supreme result of this mighty conflict
of the nations with its sending into the Beyond
these vast masses of noble youth be the development of the latent spiritual powers of man and
the recognition that death need not cause separation; that, indeed, it gives the conditions of
the closest union of spirit to spirit? Life would
be transformed; readjusted at once to a higher
plane. All its interests, and thereby its possibilities of happiness, its capacities for zest and
enjoyments, would be tremendously extended.
For the larger that one's individual world be-

comes in its potentialities of achievement, its
call to action, its unfolding of greater purposes,
the larger areas of happiness does it offer.

"Dismiss the delusion that matter is not informed with spirit, and that God knows nothing
of matter," says Archdeacon Wilberforce; "matter, incidents, material conditions, life experiences, are the spirit's media through which He
speaks to us. . . . When you blend the conscious mind with the Infinite Mind you are dwelling in the 'secret place of the Most High.' While
you are thus mentally dwelling in 'the secret
place', no sorrow can touch you, no anxiety can
fret you; you are in full communion with the
spirit beings on the other side; you are in vital
union with the Infinite Spirit."

From such communion one brings stores of
renewed energy to press on in his duties and
occupations. Humanity is on the eve of remarkable changes and transformations. The dawn-
ing recognition of powers in every individual
that link him in natural and unbroken companionship with those who have passed from the
physical realm; that make possible, by means

of this conscious recognition, the blending of effort in both worlds for the progress and up-lifting of the universal life; this general move-ment of rising to higher planes of perception is a pledge and prophecy of the most inspiring nature.

"O, Days of the Future, I believe in you!" Nor can one fail to catch on the air the wonder-ful message of the poet:

"O my brothers and sisters! It is not chaos or death. It is form, union, plan, —
It is Eternal Life, — it is Happiness!"

The messages from many of the youth who have passed on bear witness to the naturalness of the life on which they enter. There are as-pects of it that continue the aspects familiar to them here. There are new conditions resulting from the ethereal environment, about which they speculate as a man might in a foreign coun-try on confronting conditions hitherto unknown to him. "To acquire the ability to hear any-thing personally addressed to us, through any amount of space," was one thing that aroused the curiosity of the young man from whose messages several quotations have been made. What more natural?

When Doctor Graham Bell first exhibited the telephone, how eagerly people discussed this new and apparent possibility of speaking beyond the known limits of the human voice; and how incredible to the students of the invention in 1868 would have been the extensions of its service as practiced in the daily life of 1919! The speculations, and the conclusions arrived at, as revealed by Raymond Lodge; as revealed by many other of the young men; the conjectures, the assertions, the observations and inferences of all this body of youth who suddenly enter on the succeeding conditions of this endless life, form a mass of testimony that is far from unimportant. It is not an unimportant fact that the father of one of these young men who has been able (because of the coöperation of his parents) to communicate with the life here, is known as the world's greatest living scientist and one whose spiritual perceptions are so developed as to enable him to become a reliable interpreter of the nature and possibilities of this communication; one whose sympathy with other bereaved families is so great that he felt constrained to place on public record all that he felt most helpful in the messages from his own son, and thus share these with all who value them. Sir Oliver Lodge had been absolutely convinced of the reality of communion between the two worlds long before this communion had become to him so vital a matter as to its truth or fallacy. With no uncertain note he had more than once stated that he *knew* those whom we call dead could speak to us; that they are far more aware of life here than we dream; that personal communication is not only possible but that it is an assured and unquestionable fact. With the passing of his son this assurance could not but become a more vital matter to him. The comfort it has afforded is the comfort that may reach every sorrowing home. It is in the Divine Order.

Apparently these young men who in all the glow and freshness of ardent youth passed into the ethereal world so instantly, who daily faced this immediate possibility, are inevitably uplifted to the higher plane of life, whether they vanish from earth, or still remain. Life to them can never be the same again. They have stood too near to the divine realities. If they return to enter into the affairs of the present; or if they enter on the work of the next plane, they bring to bear, in either case, a new influence. Those who pass on seem to find little break in the continuity of their lives. They speak of being with their comrades the same as here. They are full of plans and interests. The special gift, or attraction, often repressed by circumstances when on earth, springs into activity in the new life.

How often, in this part of life, is it true that the one whose soul was in music has been obliged to adopt a business career; the born scientist has applied himself to agriculture or to industrial concerns. The freedom of the ethereal realm at once liberates the individual from a distasteful occupation, precisely as some suddenly fortunate circumstance in this world may set a man free from enforced labor and permit him to enter on the line for which he most cares. Mr. Lowell found his chair in Harvard a burden to his life.

He longed for the leisure demanded by his poetic gift and the freedom that would enable him to devote himself to literature. When at last this came he joyfully resigned his professorship. Similar matters of release from the distasteful occupation appear to be the experience in the ethereal life.

This assurance alone has its consolation for those in the home left desolate and bereaved. Nor is there unmixed desolation to those who find themselves initiated into the larger truth. They who understand find that understanding brings courage, trust, and joy. They who understand enter on a new and more intimate spiritual companionship with their beloved. Thus do they both give and receive a new order of happiness. For it is this gift we may still offer to the one so dear, — the gift of sympathetic comprehension of his new life. His gallant spirit heard the call, — "Whom shall I send, and who will go for us?" And in all the ardor of his divine enthusiasm he replied, "Here am I, Lord; send me!" Could the love that so tenderly enfolds him mar his new happiness with unbroken gloom and lament? Shall it not rise into perfect understanding and sympathy with the glory that has been revealed to him? The glory shall encompass life here as well as that on the higher plane. Love unites both realms, and no separation of spirit is possible. Love shares the glory and the beauty of the transfiguration. It is they who understand who shall thus enter into the gladness and the radiance which enfold and exalt the beloved in their new life and shall thus enter into the joy of the Lord.

And then?

Then, "The sun shall be no more thy light by day; neither for brightness shall the moon give light unto thee; but the Lord shall be unto thee an everlasting light, and thy God thy glory.

"Thy sun shall no more go down; neither shall thy moon withdraw itself; for the Lord shall be thine everlasting light, and the days of thy mourning shall be ended."

It is they who understand who shall enter into the realizations of the Blessed Promise. It is they who understand who shall hear, as if borne on the air, the divine assurance:

"For our light affliction, which is but for a moment, worketh for us a far more exceeding and eternal weight of glory;

"While we look not at the things which are seen, but at the things which are not seen; for the things which are seen are temporal; but the things which are not seen are Eternal!"